A Soil Owner's Manual

A Soil Owner's Manual

HOW TO RESTORE AND MAINTAIN SOIL HEALTH

By

Jon Stika

ISBN-10:1530431263
ISBN-13:978-1530431267

In memory of Mike Collins,
a tremendous advocate for educating people about soil health.

Contents

Foreword

Globally, soils are chronically unhealthy. Nationally, the over-use of chemical inputs and misguided farm programs continue to mask the chronic symptoms of dysfunctional soils. Millions of acres of uncovered, carbon-depleted soils have negatively impacted biotic, climatic, and aquatic health of our planet. The nutrients from eroded soil reach our water bodies but not our nutrient deficient human bodies. Why? We do not value our soils! We do not value what we do not understand! A majority of agricultural producers, technical personnel, and much of academia do not understand that soils are living ecosystems; much less the general public. The current national soil paradigm of soil is that of dirt; a chemistry set or a growing medium that we force and control for the sake of efficiency. Soil is not dirt; it is alive!

Jon Stika, one of the founders of the soil health movement, has laid out a path for restoring soil function. This book defines the foundational premise for viewing and understanding soils correctly. The principles of soil health are described in a simple manner that anyone can comprehend. Understanding and applying these principles to your garden, farm, or ranch can change your life. I highly recommend this book.

Ray Archuleta
Republic, Missouri

Preface

This book was written out of both compassion and frustration. Compassion for the farmers, ranchers, gardeners and others who make a living from the soil, who continue to struggle with all the symptoms of dysfunctional soil. Frustration with the status quo of agriculture, focused on yield with little or no understanding of how the soil is designed to function. There is no system of production, or soil amendment that will fix what is wrong with your soil. Only your understanding of how the soil functions will fix what ails your soil. You must become a student of what makes soil healthy. It's that simple and there are no shortcuts.

Acknowledgements

I humbly acknowledge the people that have helped me along the way on my soil health journey and putting what I have learned in this book. Ray "The Soil Guy" Archuleta, Jay Fuhrer, Dr. Kristine Nichols, Gabe Brown, Mike Zook, Dale Ferebee, Dr. Elaine Ingham, Dr. Mark Liebig, Marlyn Richter, Todd McPeak, Jerry Doan, Ray Covino, Willie Durham, Dr. Rick Haney, Ray Styer, Ken Miller, Dr. Robin "Buz" Kloot, Dave Brandt, Dr. Dwayne Beck, Dr. Jill Clapperton, Dr. Holli Kuykendall, David Lamm, Marlon Winger, Rudy Garcia, Mark Henning, Virginia Lloyd and Tanner Stika.

Special thanks to my wife, Eve, for helping me review and edit the manuscript, creating the artwork for the cover, and putting up with my endless thoughts and discussions about soil health.

Introduction

Have you ever bought or inherited a piece of machinery that did not include the owner's manual? Many of us have been in that position more than once. We put the piece of machinery into service, observe it working (or not) and make our best guess about how it is *supposed* to work. We tinker with the machine to get it to serve the purpose, but without knowing exactly how it was designed to operate, we never realize the machine's true potential and capacity to function.

Such has been the case with soil and agriculture. We tinker with the soil and see how it reacts to tillage, fertilizers, and a myriad of other inputs and make our best guesses about how to manage it. But without an owner's manual, we remain ignorant of how to manage the soil properly or understand the true capacity of the soil to function.

In recent years, soil biologists have begun to unlock the secrets of the soil and give us the makings of a soil owner's manual. What we are learning is that the soil is, in fact, a biological system run by billions of microscopic and macroscopic organisms. These organisms do their best to make the soil their home, collaborate with each other and with their partners, green plants, which live in the soil with them. The green plants take energy from the sun and feed the soil organisms, who in turn build the soil and feed the plants. Knowing this gives us a completely different perspective about how to manage the soil to make agriculture profitable, restorative, and sustainable. With a soil owner's manual guiding us in how to manage the soil as habitat for the soil organisms responsible for the majority of what is necessary for crop production, we can build soil and profitably produce crops at the same time.

The fundamental principles for improving soil health and profitability include: disturb the soil less, provide for a greater diversity of plants, maintain living roots in the soil as much of the time as possible, and keep the soil covered with plants and their residues at all times. We now possess the knowledge and technology to make this happen. Tools, such as equipment that can plant into untilled soil, multi-specie cover crops, harvesting equipment that distributes crop residues back over the soil, pesticides, fertilizers, and livestock can all be applied thoughtfully to build soil and feed the underground herd of microscopic livestock that every farmer, rancher and gardener, should have living in their soil.

Other books written about soil and soil health talk about managing dysfunctional soil by applying various amendments that run the gamut of compost, micronutrients, fertilizer, biochar or various types of microorganisms (aka "bugs in a jug"). They might acknowledge soil microorganisms as part of the picture, but miss the point of how to actually restore the soil. Instead, they have focused on how to make the soil productive while claiming certain improvements to soil health. This book is not a human-centered treatise on how to amend or control the soil to produce what *we* want, it is a soil-centric view on restoring the soil *first* and then realizing all of the crop production and environmental benefits that will follow.

There is also presently an emphasis on sustainable agriculture. Given the fact that the majority of our agricultural soils are dysfunctional, why would we want to sustain them as such? We should not be interested in sustaining dysfunctional soil. However, we should be interested in restoring the capacity of the soil to function, then sustaining that new state of full functionality.

Most of what we currently know and practice regarding crop production centers around how plants respond to inputs applied to dysfunctional soil. We have mistakenly focused on the production of plants, rather than the soil, and not looked at them together as a whole. As a result, our understanding of crop production consists of how crops will grow in dysfunctional soils and how they respond to inputs. We are just now beginning to understand how crops perform in fully functioning soil. Once our focus shifts to restoring the soil, crop production becomes a matter of managing the soil to improve its capacity to function. The soil can then better supply the needs of the plants. This is accomplished by using plants and their residues to improve the microbial habitat in the soil.

The most potent tool with which to build healthy soil is a live plant. Photosynthesis is the primary process that feeds the soil. A cropping system that creates favorable soil habitat and maximizes the flow of energy from the sun into the soil will restore soil health at the fastest rate and to the greatest extent. A soil food web that is dormant, crippled, and/or incomplete, cannot perform the tasks of building soil or cycling nutrients to feed the plants that grow in it. Plants build soil. Plants can fix what is wrong with soil. When working to restore the health of the soil we need to think about employing seeds to build the soil rather than using steel to disturb it.

Now that we *know better* regarding how the soil functions, our challenge is to *do better* and restore the soil that we have to this point degraded. With an open mind and a desire to do so, this book can help you restore your soil.

CHAPTER 1

What is Soil Health and Why Should I Care?

The terms "soil quality" and "soil health" have been used interchangeably over the past twenty years or so by those who have been looking deeper into the workings of the soil. I prefer to use the term "soil health" as it reflects the capacity of the soil to function rather than the inherent properties of a soil. For those involved in the production of crops, it would appear obvious that we would desire a soil to function to its full potential. The problem lies in that most people do not know what a healthy soil looks or acts like, nor what makes it healthy or unhealthy. Alas, we have treated our soil like dirt in our ignorance of how it functions. We have become disconnected from the soil. Anyone who cares about having food to eat and water to drink should care about the health of our soils. Those who make a living from the soil should care even more.

A simple definition of soil health is "the capacity of a soil to function" (Doran and Parkin 1994). The five main functions of soil are; maintaining biodiversity and productivity, partitioning water and solute flow, filtering and buffering, nutrient cycling, and structural support. The Merriam-Webster dictionary defines health as "the condition of being well or free from disease; or a flourishing condition". A healthy soil is one that flourishes and allows the plants that grow in it to flourish as well.

There are three main classes of soil properties; physical, chemical, and biological. The physical and chemical properties of soil are fairly well understood and relatively easy to measure. They include the proportions of sand, silt and clay, amount of acidity, alkalinity, nitrogen, phosphorous, potassium and the like. It is the biological properties of the soil that remain the least understood and most dynamic, yet have the greatest potential to impact soil function and plant productivity.

It would be reasonable and logical to say that if the soil is not performing all of the functions we expect of it then the rest of the environment, including the water, air, plants, and animals, will suffer. This is the crux of the matter when we look at air and water quality in a world where most agricultural soils are fundamentally dysfunctional. We see evidence of soils failing to perform basic functions of partitioning water and solute flow (resulting in floods and droughts), filtering, buffering and nutrient cycling (resulting in excessive nitrogen and phosphorus in surface and ground waters such as the perennial dead zone in the Gulf of Mexico). There are few natural processes critical to our survival that do not reside in, or pass through, the soil.

If it is true that most of our agricultural soils are dysfunctional, then we must understand the impact that our management of the soil has on the life in the soil and its capacity to function. It is the living organisms (both plant and animal) in the soil that perform many of the processes that keep the soil functioning. All of the organisms and their by-products are collectively known as "organic matter", which is sometimes referred to as the "heart of the soil". In her Vimeo presentation (https://vimeo.com/21310772) "The Biology of the Soil", Dr. Kristine Nichols notes that soil organic matter (including living organisms) controls 90% of soil functions. It is this biology that is responsible for the majority of what the soil does, including produce crops. It is the biology that builds the physical structure of the soil and regulates the chemical processes in the soil. Without *biology*, soil is simply *geology*.

Like all the muscles, valves, and arteries in our own heart, it's all the small moving parts in the soil that make it run. To keep the heart or our soil running, we must understand how all the parts work together and how to keep them healthy and functioning to be productive and sustainable.

The Soil Habitat

"Habitat" is the physical location where an organism is found. Soil microorganisms (sometimes called "microbes") are microscopic, and inhabit some very tiny places in the soil. They need the space in pores and cracks between soil aggregates that are between .5 and 10 microns in size to live. Soil aggregates are the lumps, clumps and crumbs of sand, silt, and clay, stuck together by sticky substances produced by plants and soil microorganisms. More about aggregates later. Any space less than .5 microns is too small for even a bacteria to squeeze into, and anything larger than 10 microns is typically too big and dry. Bacteria can inhabit about 30% of the spaces in a clayey soil but only 13% of the pores in a sandy soil. Ideally, soil should be about half solid and half spaces, with half of the spaces containing water and half containing air.

A soil's water holding capacity and temperature are also critical to life in the soil. Microorganisms are most active between 50% and 75% soil moisture content (by weight) and at a temperature of between 40°F and 80°F. For every 10°F temperature increase between 40°F and 80°F, the activity of soil microbes will approximately double if adequate moisture is present. Because air doesn't circulate as well below ground as above, just a 1% increase in the concentration of oxygen in the soil will also significantly increase the activity of soil microbes. Allowing too much oxygen to be available to soil microorganisms can spell trouble for soil health. This is akin to stirring a fire to inject oxygen to increase the rate that fuel will burn. More about that in Chapter 2.

Therefore, we must change the way we perceive the soil in order to change the way we manage the soil. Humans have been tillers of the soil for roughly 7,000 years. It has been our erroneous perception of the soil and failed attempts to control it that has resulted in its degradation. We must look at ourselves as *part of* the environment rather than something *separate from* the environment. It is that disconnectedness and desire for control that defines both our historic, as well as our current approach, to agriculture.

Agriculture's Impact and the Need for a New Paradigm

Because our modern approach to agriculture is focused on the production of crops in monocultures (a single species of plant grown at one time), we have placed ourselves on an expensive treadmill which is at the mercy of the cost of inputs for that system of production. Instead of maintaining and relying on healthy soil, we have instead relied on plant breeding, fertilizers, tillage and pesticides, as the basis for agricultural production. Current agronomic practices appear to be focused on supplying a myriad of inputs hoping for an output without a true understanding of what lies between. It has been estimated that annual agricultural inputs of nitrogen and phosphorous on U.S. agricultural lands exceeds the nitrogen and phosphorous contained in harvested crop exports by 40-60% (Drinkwater & Snapp, 2007). These excess nutrients are creating tremendous water quality problems across the country, and may continue to do so for decades to come (Van Meter et. al., 2016).

However, a new approach is emerging that looks beyond meeting the needs of a single species of plant to include the relationship that plants should have with the soil and soil with plants.

A striking example of this new approach to agriculture is the Brown Ranch, operated by Gabe Brown and his son Paul, near Bismarck, North Dakota. The Browns' have changed their model of production from a focus on crop yield to one of restoring soil health by integrating the production of cattle and crops together. After a series of setbacks while attempting to follow the standard spring wheat production model, Gabe and Paul realized that by focusing on the soil first, their costs of production fell and their ranch's resiliency to the extremes of weather increased. Extreme dry years and wet years did not significantly affect their ability to be both productive and profitable. During a particularly dry year, one of the Browns' neighbors implored them to remove bales of hay from a field that was visible from the highway because he was trying to make a case for a drought declaration and assistance from the government. Because of the capacity of the Browns' soil to capture and store water, they had much greater production than anyone else in the vicinity. This begs the question, was the drought caused by lower than normal precipitation in the area, or the lack of soil health and the capacity of the soil to store and supply water to the plants?

Our conventional approach to crop production is akin to delivering steel, rubber, and plastic to a factory and hoping lots of good automobiles come out of the factory, without knowing anything about the workers or what they need to do their job. If we remain unaware of the process of building cars, what might come out of the factory would be left mostly to luck. So, if the soil is the factory of the farm, we need to understand who works there and what those workers need to be productive. Most of us would have to admit we don't know much about the workings of the soil factory.

Green plants have been interacting with a huge array of organisms in the soil in many different ways for a very long time. Plants and soil organisms have established these age-old relationships to assure each other's survival. The microscopic life in the soil obviously knows how to make and maintain soil. This soil building process has been going on for many years with little or no human intervention. Our crop production systems need to mimic the conditions that optimize that soil building process again.

A farmer's objective should be to use plants to efficiently harvest water and sunlight. The role of the plants is to use water and sunlight to manufacture carbohydrates, fat, and protein. The soil's role is to feed and water the plants. The mistake we make in agronomy is attempting to feed and water the plants directly, as if the soil were in the way, instead of recognizing that the soil *is the way* to get the job done.

5

To be successful, farmers, ranchers and gardeners need to understand what it takes to care for the soil so plants can make the greatest use of water and sunlight; maximizing soil health to allow the soil and plants to interact at their best. Doing this, should result in the greatest crop yield for the least cost. Mike Zook, from Beach, North Dakoa is one farmer who has been practicing regenerative agriculture for some time. I had the privilege of interacting with Mike for years and watched as he improved both the health of his soil and his bottom line in Golden Valley County, North Dakota.

The Fallacy of Feeding the Plants

In a paper titled; "Nitrogen: The Double-Edged Sword", Dr. Christine Jones (Jones 2008) describes how nitrogen fertilizer affects soil biology. If the entire amount of nitrogen that a crop needs is added to the soil at planting time, it will often suppress the association soil microbes have with plants. Plants and soil microbes will use the applied nitrogen independent of each other instead of relying on each other, delaying many vital associations between them.

With access to excess nitrogen, soil microbes will feed on the carbon stored in soil organic matter, resulting in the liberation of plant nutrients, including nitrogen. In a natural state, plants trade sugar to soil microbes in exchange for nutrients. When we apply excessive nitrogen fertilizer to the soil, the plant-microbe relationship is bypassed early in the growing season. The plants will use the rapidly available nitrogen from fertilizer instead of relying on soil organisms to supply it to them from naturally occurring sources. Later, when the excess nitrogen is gone, and the plants neglected to make suitable associations with their partners in the soil, the crop suffers as it spends time and energy trying to build an association with the soil microbes that it shunned earlier. As the growing season progresses and the plants require additional nitrogen from soil microbes, they will exude sugars out of their roots to attract and feed microbes. This is the stage of growth when plants typically should be sending energy (sugars) to the seed head to produce grain. Eventually the plant-microbe association will be created, but often too late to meet the needs of the crop at a critical stage of grain production.

The plants, in effect, become nitrogen addicts, dependent on an artificial supply of nitrogen in lieu of maintaining a healthy relationship with the rest of the soil community. This not only results in poor late-season crop performance, but can result in some of the nitrogen leaching below the root zone before it can be acquired by soil microbes or plants. This phenomenon occurs whenever any form of nitrogen fertilizer, organic or synthetic, is over-applied to the soil system. It is currently understood by most agronomists and farmers that there are only two forms of nitrogen available that plants are able to access from the soil, nitrate and ammonium. However, it has recently been determined that plants can also utilize amino acids and proteins found in the soil as sources of nitrogen (Lonhienne et. al. 2008). This opens up a whole new understanding of how nitrogen is not only found in the soil, but how it is accessed by plants. Amino acids and proteins are products of soil organisms, not part of applied fertilizers. An active and functioning SFW can do a great deal more to supply nitrogen to plants than we thought! This reinforces our understanding of how the soil functions as a biological system, instead of simply a chemical system.

I had these plant-nitrogen interactions explained to me by Dr. Kristine Nichols, Rodale Institute Chief Scientist, formerly a Research Soil Microbiologist with USDA's Agricultural Research Service. I was amazed at the collaboration that occurs between plants and the Soil Food Web (SFW). This understanding shattered much of what I knew about modern agronomy. More about the SFW later.

The plant-SFW interaction just described, regarding nitrogen, also occurs to a lesser extent with phosphorous. Excessive phosphorous fertilization resulting in more than 15ppm in the surface three inches of soil can inhibit the association of AMF (Arbuscular Mycorrhizal Fungi) with plant roots. This can limit a crops' ability to access water and nutrients later in the growing season when a greater association with AMF would help bring more water and nutrients to the plant than its own root system could acting alone (Smith et. Al. 2001). More about AMF, or more simply mycorrhizal fungi, in Chapter 4.

The Suitability of Feeding the Soil

If feeding plants directly is folly, then we must learn to feed the soil so it can feed the plants. More precisely, we must learn to feed what lives in the soil. We must understand the soil factory and its workers.

The soil factory exists in and between soil aggregates, which is often referred to as 'pore space'. Soil aggregates are the crumbs and clods of soil held together by the sticky substances and filaments of the microscopic life in the soil. The workers in the soil not only run the carbon, nitrogen, and phosphorous cycles, they build and maintain the factory itself. Tillage crushes soil aggregates, damaging the soil factory. Therefore, it would seem that the first order of business is to limit this damage as much as possible. Tillage cannot create stable soil aggregates, only break them. There is no tillage operation that directly benefits soil structure and function (Brady & Weil 2002).

Crop production without tillage benefits the soil by not disturbing it any more than necessary to accomplish the placement of seed in the soil to germinate and grow. The only soil disturbance necessary should be a function of the size of the seed that needs to be placed in the soil and the spacing of the rows to achieve a suitable stand of crop. A single-disk opener drill making a one-inch slice into the soil every 7-8 inches should be all that is necessary to accomplish seed placement for most crops. In this way, the soil factory and its workers are not disturbed any more than necessary to accomplish the planting process. The technology now exists to plant seeds and transplants in this manner, directly into an undisturbed soil.

The methods of planting without tillage have been perfected by a group of dedicated farmers known as the Northern Prairies Ag Innovation Alliance (formerly the Manitoba-North Dakota Zero Tillage Farmers Association). The original association was formed back in 1982. I learned a great deal from them about zero tillage and soil health during the time I served as an advisor to them during the 1990s. The Alliance has published several works, including three manuals on zero tillage farming and soil health, in addition to organizing many workshops to help farmers learn their methods (see appendix).

The next order of business to expedite a productive soil factory is to create an environment that facilitates the greatest diversity of workers in the soil. A very diverse workforce is capable of handling all the tasks of the water, carbon, nitrogen, and phosphorous cycles with built-in redundancy to negotiate a variety of environmental conditions. If workers who specialize in a particular part of the process are not present, that part of the process will lag behind and slow overall soil function. The key to soil workforce diversity is to have the greatest diversity of partners (plants) and food sources available to those workers. (Zak et. al. 2003).

The New Paradigm

In order to help the soil feed the plants, less disturbance, more diversity, and a carefully considered input of fertilizer can facilitate soil function by recognizing the soil as a factory full of working organisms. As soil function increases, so will the ability of the soil to power itself and feed the plants that grow in it. This should allow us to produce more with less, as the soil functions at ever-higher levels. Like an injured person who is rehabilitated to run a marathon, we can rehabilitate our soils as well. We have done such great injury to our soils over the years that we have become accustomed to their state of dysfunction and accept that degraded condition as normal. As we begin to rebuild our soil factory and repopulate and rehabilitate the workers, we are getting a glimpse of the potential output our soils are capable of generating. We must recognize that the sustainability of agriculture ultimately depends on the capacity of the soil factory and the workers that labor within it.

In my travels as an agronomist and soil scientist in western North Dakota, I often observed sloping fields that had visible rills and gullies in the soil even though there were reasonably good crops growing on them. I also observed many other fields with similar slopes and soils that did not show any visible signs of rill or gully erosion. After some checking, I found that the fields with rills and gullies were those farmed using a considerable amount of tillage. By contrast, the fields without symptoms of rills and gullies had not been tilled for the preceding several years. It dawned on me that in the past, we used to think that soil erosion on slopes was inevitable during significant rain events because of the water that would run off from the soil. The truth is that the soils experiencing runoff had poor aggregate stability and thus, were dysfunctional in their capacity to infiltrate water. More about aggregate stability later.

Now that the majority of the crop land in western North Dakota is no longer tilled and has at least a fair degree of crop rotation diversity, the soils are functioning better, with more water infiltrating the soil instead of running off. Soil erosion caused by water runoff was not a problem, but a symptom of soil dysfunction. In the past, we were struggling with collapsed, dysfunctional soils and trying to figure out how to deal with all the water that was running off the soil from those fields. Those soils did not have a runoff problem, they had an infiltration problem. The water ran off because it did not infiltrate. This was due to a lack of aggregate stability at the soil surface. Now that we realize that erosion is a symptom of a soil that is not functioning properly, we can deliver the message to producers that they can restore their soil and get it to function again; eliminating the symptom of erosion.

There have been many folks that have attempted to understand the reasons why farmers have been reluctant to apply conservation practices to their land. I believe it is because the message of "you have erosion, you need to control it and it will cost you money" was not a positive one. Conservationists admitted that they could only help farmers slow the decline of their soil from the effects of erosion and catch the water and sediment as it tried to leave the field. This was not a very hopeful message and did not usually elicit positive responses from producers.

In contrast, a message of restoring soil health is, "We can help you rebuild your soil and get it to function again while reducing your input costs." This is a message of hope. This is a message of restoration and true sustainability. Producers are responding positively to this message and are applying conservation practices as tools to pursue the goal of building soil health rather than simply to stem erosion. The primary obstacle to resolving the symptoms of water runoff and erosion is a lack of understanding of how the soil functions. Armed with this new understanding, we can now eliminate resource symptoms that are the result of dysfunctional soil.

The Soil Paradigm Shift

The Merriam Webster dictionary defines a **paradigm** as: *a philosophical and theoretical framework of a scientific school or discipline within which theories, laws, and generalizations and the experiments performed in support of them are formulated; broadly: a philosophical or theoretical framework of any kind.*

Most people's paradigm of soil is typically characterized as "dirt". When in reality, soil is a living ecosystem that includes billions of organisms of untold species living in concert with each other and living plants. The soil paradigm shift that needs to occur, is one of people changing the way they think about soil from that of "dirt", to thinking about it as a "living ecosystem". Soil is alive!

Producers need to ask themselves; "What functions do I expect my soil to perform?" When I have asked this question of farmers and others during workshops, I am typically met with blank stares. This is not because these folks were not intelligent and thoughtful people. It is because no one had ever asked them that question before! Nevertheless, this fundamental question must be examined if any farmer, rancher or gardener expects to be profitable and sustainable.

The first response from most people when asked, what functions they expect their soil to perform is, "To grow crops." When asked what crops need in order to grow, they quickly answer, "Water, nutrients, sunlight and air." When asked if they know how soil must behave for water to infiltrate into it, their answer typically includes the belief that tillage opens up the soil to allow water to infiltrate. In fact, tillage only briefly opens the soil while simultaneously *reducing* the capacity of the soil to infiltrate water later by destroying soil aggregates and the biologic glues that hold soil aggregates together. This results in the collapse and crusting of the soil with the next rain or application of irrigation water. The more the soil is disrupted with tillage in an effort to increase water infiltration, the more rapidly it will collapse and seal the next time the soil becomes wet.

When asked if they know how plant nutrients become available to plant roots, producer's answers typically include the belief that fertilizer must be added to the soil, where the fertilizer dissolves in soil water and the plants take the nutrients in. In fact, 90% of the nutrients taken up by plant roots are cycled through a soil organism before becoming plant-available. Virtually everything plants need is supplied by the soil organisms that live in collaboration with each living plant (Lavelle & Spain 2005). Less than a third of the nitrogen fertilizer applied to a field ends up *in* the plants grown there (Stevens, Hoeft & Mulvaney 2003). The rest is retained by some other form of life in the soil, volatizes into the atmosphere, runs off the field or leaches down below the root zone in the soil with the movement of water. Most analytic soil testing and fertilizer prescriptions are based on the response in crop production of plants grown in dysfunctional soils. The methods and prescriptions work quite well; *for dysfunctional soils* (Laboski et. al. 2006). This should come as no surprise, since most agricultural soils in the U.S. do not cycle nutrients very well, so the corresponding methods of testing and prescribing fertilizer application have evolved accordingly.

Water infiltration and nutrient cycling are just two basic examples of what we now understand are processes that are driven by the organisms living in the soil. This change in understanding of how the soil works as a biological system is a major paradigm shift for almost everyone in agriculture. Armed with this new understanding of soil function, producers can reduce and eliminate the symptoms of erosion, runoff, nutrient leaching, drought, and poor crop performance to become truly sustainable.

The bottom line is that the plant available water in the soil becomes plant available because soil microorganisms made the soil aggregates that allow the water to infiltrate and be stored in the soil. It is also soil microorganisms that cycle and make the vast majority of nutrients available to plants.

If asked, any producer will tell you that they expect their soil to grow profitable crops by supplying water and nutrients to their crops. *What many folks don't realize is that these two basic expectations of soil function (water and nutrient supply) are biologically driven.* Keep the soil microorganisms happy and the system runs at peak efficiency. A more efficient system will be a more profitable system.

CHAPTER 2

What's Wrong With My Soil Anyway?

A s already mentioned, most agricultural, urban, lawn and garden soils in the United States are in some degree dysfunctional. This can be easily determined by performing the simple tests listed in Chapter 7. Most soils that have been altered from their native condition suffer from some combination of too much disturbance, too little plant diversity, not enough time hosting living roots and a lack of soil cover. I have witnessed millions of acres of dysfunctional soil during my time and travels as a soil mapper, conservationist and soil health instructor. I have encountered degraded soil on conventional farms, organic farms, no-till farms, vineyards, orchards, hay land, pastures, and native rangeland, all due to a lack of understanding of how soil functions.

A typical farm field is often tilled multiple times each year, is planted to one or, at best, a few different plants at a time or in rotation, has periods of time where no living plants inhabit the soil, and is often left uncovered, at some time during the year. A typical lawn has its plants mowed short, clippings bagged and exported (often to a landfill), watered too frequently, over-fertilized, and often consists of only one or two species of plants. We have come to accept dysfunctional soils because they have been dysfunctional for generations. It is time to realize that we have not understood how soil is designed to function and that our lack of understanding has resulted in degradation of our agricultural and urban soils.

Soil Disturbance

There are three types of disturbance that can impact how the soil functions: physical, chemical and biological. Physical disturbance includes anything that fractures, compacts or otherwise moves soil from one position to another. This can occur when the soil is tilled or has some sort of traffic (tractors, trucks, livestock, etc.) move across it. Chemical disturbance includes anything that we do that changes the pH (acidity or alkalinity), salinity (salt content), or nutrient content (nitrogen, phosphorous, potassium, etc.) of the soil. The addition of pesticides may also be a chemical disturbance to the soil. Anything we do that significantly alters the nutrient content or biologic toxicity of the soil is a chemical disturbance in the world of soil microorganisms. Biological disturbance typically occurs in the form of a lack of plant diversity, or the degradation of the plants themselves (such as overgrazing). A soil inhabited by a single species of plant will not feed the soil organisms a proper diet.

We must be aware of what we add to the soil and how it will affect the organisms that live there. There is a significant lack of information regarding the effect that pesticides may have on soil organisms. Some pesticides are lethal to certain soil organisms and other pesticides may be broken down by soil organisms and used as a source of energy or nutrition. Much is still unknown about how many of the chemical compounds we apply to the soil affect the organisms that live there. We need to be aware that applying such things as fungicides or fumigants can do significant damage to the life in the soil. Biological disturbance due to the lack of plant diversity and overgrazing will be discussed later in this book. The primary and most destructive disturbance we do to the soil is physical.

A case in point of the degradation that has occurred to our soils is the decline in the amount of carbon in soil organic matter in our agricultural soils. Research by the United States Department of Agriculture (USDA), Agricultural Research Service researchers Don Reicosky and Mike Lindstrom has shown that carbon leaves the soil in the form of carbon dioxide very rapidly when the soil is torn open by tillage (Reicosky et. al. 1993). When the soil is opened up by tillage, carbon dioxide leaves the soil much the same way carbon dioxide leaves a can of carbonated beverage; from higher concentration in the soil to lower concentration in the atmosphere above the soil. Over the course of several days or weeks, copiotrophic bacteria (bacteria that feed and reproduce quickly) use the additional oxygen added to the soil by tillage to voraciously consume the most fragile fraction of soil organic matter, the biologically produced glues that hold soil aggregates together.

Biologically produced glues (such as glomalin produced by symbiotic mycorrhizal fungi that connect with living plant roots) need to be continuously produced in the soil because bacteria and other soil organisms are continuously consuming them. Glomalin is a glyco-protein, a sugar-protein that mycorrhizae produce as a coating for the outer surfaces of their hyphae (root-like structures) to help prevent loss of water and plant nutrients (Nichols 2008). Glomalin is thus a practical foodstuff for smaller, simpler organisms in the soil. If bacteria are allowed to consume glomalin faster than it can be produced by mycorrhizal fungi, soil aggregates will more easily disintegrate when they become wet.

When the soil is tilled, aggregates are cracked open and allow bacteria access to a variety of carbon-based food (including glomalin) that was previously not accessible to them. At the same time, the soil is aerated with atmospheric oxygen. This combination is akin to throwing small twigs on a fire and stirring it about, causing the fire to flare up and burn the available carbon quickly. It is this process that has been responsible for reducing the native concentrations of organic matter in soils to the low levels we see today.

Research conducted in the Red River Valley of North Dakota by David Hopkins and Brandon Montgomery of North Dakota State University showed this decline in soil organic matter quite clearly. They visited the exact locations of several soils examined as part of soil surveys during the 1960s by the USDA. One particular soil in Walsh County had 34 inches of soil above the C horizon (the original material left behind in the area by glacial Lake Agassiz). The soil was examined again in 2014 and was found to have only 15 inches of soil above the C horizon; a loss of 19 inches of soil in roughly 50 years. The soil examined in 2014 did not contain the same quantity or quality of organic matter that it did at the previous investigation in the 1960s. Through the mixing action of tillage and the application of synthetic fertilizer to support crop yields, the losses to the soil were not often noticeable to the casual observer.

Many of the agricultural soils in Walsh County currently contain about three percent organic matter. They originally contained nearly eight percent organic matter. Some of the original organic matter soil blew away with the wind as it was often left in a tilled, unprotected condition. The balance of the organic carbon left the field as *aerobic erosion*, through elevated microbial respiration facilitated by excessive tillage. The soil organisms literally consumed the soil organic matter under the conditions created by tillage. The organic carbon left the soil as carbon dioxide; the product of elevated microbial respiration brought on by aeration of the soil by tillage. This is the case with many agricultural soils. Soil is lost as an invisible gas (carbon dioxide) as well as by the soil physically washing or blowing away.

Not only has excessive tillage been to blame for the reduction in *quantity* of soil organic matter, it is also at fault for reducing the *quality* of soil organic matter. Soil organisms and their by-products, such as glomalin, define the *quality* of soil organic matter vital to the capacity of the soil to function. It is the fraction of organic matter that glues soil together into aggregates that is the most fragile and easily lost. Thus, the *quality* of soil organic matter often declines first, reducing the capacity of the soil to infiltrate water, followed by a reduction of the *quantity* of soil organic matter as water runs off the soil, carrying soil with it.

Physical soil disturbance, typically in the form of tillage, breaks down soil aggregates and facilitates the consumption of biologic glues that were supposed to hold the aggregates together. This leaves the soil in a collapsed state, prone to further collapse the next time it rains. Since the soil gets wet from rainfall or irrigation at the surface first, it will collapse at the surface first, rapidly sealing off pathways that existed when the aggregates were intact. Thus, a tilled soil will seal faster and drastically retard water infiltration resulting in ponding of water in flat areas or runoff of water in sloping areas. Agricultural soils do not have a runoff or erosion problem, they have an *aggregate stability problem*, indicating poor soil health by the reduced capacity of the soil to perform the basic function of infiltrating, filtering, and storing water. This bears repeating: *Soil erosion is not a problem. It is a symptom of unhealthy, dysfunctional soil.* Much time and money has been spent addressing the symptom of erosion instead of the problem of soil dysfunction.

Simply put, plants (equipped with green chlorophyll) use solar energy, carbon dioxide from the air, water, and nutrients, from the soil, to make simple sugars (carbohydrates), proteins and fats that make up the various parts of plants. When these plants die and decompose, the carbohydrates and proteins get converted from plant material, to soil microbe bodies, and then to organic matter (the stuff that makes topsoil dark). This organic matter is responsible for the majority of the good things soil does, like holding and releasing water and nutrients for plant growth, resisting erosion, filtering water, etc. The reverse of photosynthesis occurs when we "stir the fire" by tilling the soil and introducing lots of air (oxygen) so the microbes can work overtime eating the soil organic matter at an artificially rapid rate. A classic example of this process occurs when soil is tilled to control weeds during fallow periods between crops.

Soil tilled repeatedly during times when the soil is warm and moist can result in very high rates of microbial activity, converting soil organic matter to nitrate-nitrogen and carbon dioxide. The crop response we see after a tilled period of fallow is due to the nitrogen that soil microbes converted from organic matter to nitrate, not from an increase in soil moisture. Tilled fallow practiced in drier parts of the country is only about 5% efficient at capturing and storing moisture for the next crop because the soil is often left uncovered and subject to extreme heating and drying (Bauer & Conlon 1974).

Some years ago, I visited with a farmer who told me that he sought out land that had a history of such fallowing practices because he knew there would be nitrate nitrogen that had leached below the root zone of wheat. This was in an area of North Dakota with a total annual precipitation of 17" of water. He would rent such land and plant it to sunflower, which rooted much deeper into the soil than wheat. In this way he would take advantage of all the "free" nitrogen that came with the land that could be tapped by the sunflower crop.

Understanding this part of the carbon cycle should beg the question: Is mining soil organic matter for nitrogen sustainable? Given the fact that many soils in North Dakota were originally more than 4% organic matter and are now 2% or 3% (sometimes with little or no history of physical erosion), should indicate the answer is no. This reduction in soil organic matter content is true in most parts of the country with a history of tillage.

In the days before commercially available fertilizers, native soil organic matter was the most economical source of nitrogen available to produce a crop. Fallowing with tillage was an obvious choice to supply nitrogen for higher yields. However, now that we know how to store and cycle carbon and nitrogen with plants, we can restore the organic matter content of our soils instead of depleting them in a quest for nitrogen. A better understanding of soil biology can help us understand a number of things we see in the field and are at a loss to explain. As mentioned previously, soil organic matter is the heart of the soil. Damage or remove the heart and the patient will be in serious trouble.

CHAPTER 3

How Is Healthy Soil Supposed to Function?

A healthy soil is alive and should look, smell, feel, and function as such. Casual observation of a healthy soil should reveal a dark color from organic matter, arthropods (insects and their cousins) and hopefully, earthworms. The soil should have stable aggregates, yet be easy to dig into with hand or shovel. Plant roots should grow straight down and throughout the soil. There should be many pores of various sizes with little or no evidence of compacted horizontal layering. You should be able to easily break apart healthy soil with your fingers. Compacted and/or crusted soil indicates poor aggregate stability and poor soil health.

A healthy soil should smell slightly sweet with the distinct aroma of geosmin. Geosmin is a by-product of actinobacteria, a microscopic filamentous bacteria that performs some of the same functions of decomposition in the soil as fungi. A soil that smells like rotten eggs (Hydrogen Sulfide) indicates a soil dominated by anaerobic organisms (healthy soil should be aerobic – containing oxygen). A soil with a metallic or kitchen-cleanser (i.e. Ajax® or Comet®) aroma is often dominated by bacteria and out of biological balance. The microbiological community of the soil should be complex, diverse, balanced, redundant, and include bacteria, fungi, actinobacteria, protozoa, nematodes, arthropods, earthworms, and many others. People may think you are strange for smelling the soil, but it is an assessment I make regularly to give me an idea of how complete the soil food web is in a soil. See the appendix for a lab where you can have soil analyzed for the particular presence and populations of soil organisms.

A healthy soil should perform the basic roles of water cycling, nutrient cycling, and physical support. It should perform all of these while supporting a diversity of life that will facilitate its continued capacity to function. Find a location in your area where the soil has never been cleared of native vegetation or tilled. Look at that soil in its native condition to get an idea what a functioning soil should look and act like. A large part of the problem with current soil management is that we regard human-degraded, dysfunctional soil as a normal condition. Many folks involved in modern agriculture have not observed a fully functioning soil and therefore have difficulty comprehending how it should look, feel, smell and function.

Water Cycling

The functional capacity of the hydrologic (water) cycle on most agricultural soils on Earth is currently rather poor. Evidence runoff, erosion, floods and droughts; all symptoms of soil not properly performing the function of capturing, storing, supplying, and filtering water as part of a viable hydrologic cycle. Runoff, erosion, floods and droughts often make the news as natural disasters beyond our control. While it is true that naturally occurring extreme rain events can sometimes produce catastrophic runoff, usually the true disaster is likely dysfunctional soil. This is the result of the collapse of soil structural stability at the most basic level, the soil aggregates. Soil aggregates are the little crumbs, clumps, and blocks that give soil the capacity to act as a living, porous medium. 80% of the organic matter in the soil is held within soil aggregates less than two millimeters in size. Smaller aggregates are typically more stable than larger aggregates and affect water infiltration into the soil accordingly. Thus, the degree and stability of soil aggregation may not be apparent upon casual observation (Hoorman & Islam 2010). Without water-stable soil aggregates at the soil surface and pore space in the soil below, the soil will not allow water to infiltrate as it should. Yes, it really is that simple. The severity of symptoms of water runoff, soil erosion, floods and droughts can all be traced to this simple concept of soil dysfunction.

Soil managed as biological habitat will create and maintain stable soil aggregates and allow water to infiltrate and permeate the soil, making it available to plants and buffering the flow of water into surface waters and groundwater. Soil aggregates are built by soil microorganisms and plants, not by tillage. Tillage can only degrade and disintegrate soil aggregates. Tillage induced aerobic erosion of soil organic matter results in the loss of the most fragile fraction of soil organic matter, the organic glues that keep soil aggregates water stable. This can be observed directly when rain falls on previously dry soil. If soil aggregates are not water-stable, the soil will collapse first at the surface, filling soil openings with soil separates of sand, silt and clay and restricting infiltration of water into the soil. After the rain ends, the surface of a dysfunctional soil dries into a crust. This crust is the collapsed remains of former soil aggregates. The resulting crust retards air and water movement into the soil. This is the downward spiral of worsening soil health that has been practiced on most agricultural land for thousands of years.

Nutrient Cycling

Soil and plants capture (immobilize) and subsequently release (mineralize) plant nutrients. Plants feed the microorganisms in the soil through photo-synthesis and the soil microorganisms feed and water the plants. If the soil is managed as a biological system, this results in few, if any, losses of nitrogen or phosphorous to surface water, ground water or air. This is often referred to as biogeochemical nutrient cycling. In a naturally and fully functioning soil nutrient cycle, plant nutrients are held for short or long periods of time in the cells of plants or soil organisms. In this way, the nutrients rarely leave the field, but are cycled and supplied to the growing plants. That is the way nutrients were designed to cycle in the soil.

In most present-day agricultural systems, primary crop nutrients are supplied from outside sources in various synthetic and natural forms. This is necessary because most soils no longer include the biological capacity that allow the soil to feed the plants. This is the basis for modern soil crop nutrient testing and prescriptions. Present agronomic soil testing and prescribing methods are well researched and accurate for dysfunctional soil. Methods of assessing the capacity of functioning soil to supply crop nutrients are currently under development and are based on the number and diversity of organisms in the soil, their habitat, and the availability of food. More about nutrient cycling later.

Once we recognize soil organisms as the drivers of soil health, we understand that the most important element in soil nutrient cycling is not nitrogen, phosphorous or potassium, but *carbon*. Carbon is the currency of the soil. It feeds the organisms that comprise the soil food web so they can fix, decompose, acquire, and cycle essential plant nutrients. Carbon enters the soil economy through plants or other photosynthesizing organisms that possess chlorophyll. This is why it is important to have living plants occupy the soil as much of the time as possible; to keep the underground "herd" of soil organisms fed. Soil organisms that are well fed keep plant nutrient cycling active and balanced. Plant nutrients are acquired and supplied to plants in a relatively tight cycle with few "leaks" to surface-water, groundwater or air. A healthy, functioning soil is not nutrient-leaky. Nutrients cycle from plants to soil organisms and back again. 30%-40% of nitrogen in the soil is in the form of amino acids and amino sugars which are all compounds assembled by organisms that live in the soil (Hoorman & Islam 2010).

Blooms of algae, dead zones in surface waters, and nitrate pollution of groundwater, are caused primarily by the mistaken belief that the soil is exclusively a physical and chemical system. Witness the significant concentrations of nitrate nitrogen in the Ogallala aquifer of the Great Plains. It is this lack of understanding of the soil as a biological system that is the cause of most environmental pollution associated with agriculture. Another example of this is occurring in Des Moines, Iowa where the city is suing three counties in the Raccoon River watershed (one of Des Moines municipal water sources) for nitrate pollution. The removal of nitrates is costing the city a large sum of money to supply water that is safe for the people of Des Moines to drink.

Soil organisms control the pH and availability of nutrients in the soil and work in concert with plants to protect and supply the roots with these nutrients (Walker et. al. 2003). Nitrogen, phosphorous, potassium and other plant nutrients do not lie around in the soil in excessive quantities or in soluble forms for very long. Nutrient cycles in healthy soil are kept tight and efficient, not loose and wasteful. Nitrogen in particular is held in the cells of soil organisms or plants except for a brief transfer between them. Thus, healthy soils are economical, efficient, and environmentally friendly.

Physical Support

The soil must maintain its structural integrity (stable soil aggregates) to allow for water infiltration, root penetration, air exchange, and to support traffic by equipment or animals over the soil surface. Stable soil aggregates are the key building block of functioning soil. Soil aggregates can only be built by soil organisms and plants. Soil aggregates can by assessed for water stability by simply drying them and then immersing them in water to see how long they hold together as they become wet. This is often referred to as the slake test or aggregate stability test. More about this test in Chapter 7.

While it is true that clay lends some structural integrity to soil aggregates, it is the renewable biological glues that are responsible for creating soil aggregates in a relatively water-stable form (Pikul et Al. 2009). Soil structure extends from the soil surface down to several feet in depth. It is the stability of soil structure at the surface that is critical for air and water movement and physical support.

Biodiversity

Biodiversity simply means a diversity of life. A very simple concept, but often under-appreciated. It is a basic tenet of ecology (the study of the interaction of organisms with their environment) that diversity = stability. Stability provides resistance to drastic change and resilience to respond to changes that do occur. Biodiversity of life in the soil is what is responsible for facilitating the majority of functions we expect the soil to perform. Without adequate numbers and diversity of organisms in the soil, the soil collapses into a condition of dysfunction.

A diverse soil food web includes a great deal of redundancy so the soil can continue to function over a wide range of conditions. Biodiversity supplies willing workers to perform nutrient cycling and aggregate building in the soil. A soil that lacks biodiversity will have a greatly reduced capacity to function because members that perform certain tasks are lacking or unable to do their work under certain moisture or temperature conditions. The soil should continue to perform the functions we expect of it under the greatest range of conditions possible. If it won't, then it is akin to a hired hand who lays down under a shade tree when the weather is hot, or sits in the house when the weather is cold. The person is still present on the farm, but is not getting any useful work completed. We need a diversity of workers that will work under the greatest range of conditions possible to keep the soil functioning.

CHAPTER 4

Biology of the Soil

he soil is a living system. It is not an inert pile of dirt described only by its physical and chemical properties. *Soil is alive.* Soil microbiologists have determined that roughly 90% of the functions we expect the soil to perform are biologically driven. The number of organisms in a handful of healthy soil far exceeds the population of humans on Earth. To underestimate or ignore the biological aspect of the soil is the reason our ground and surface waters are polluted and few people can actually make a living off of the land without taking off-farm work or subsidies. Modern agriculture has left our farms and farmers in a sorry state, all for a lack of understanding how the soil functions! Appreciating soil biology and all that it does is essential to improving soil health and becoming an economically and environmentally sustainable producer.

The Soil Food Web

We are all familiar with food chains where a small fish is eaten by a larger fish which is eaten by one yet larger again. The connections of who eats who and who collaborates with who in the soil is often described as the Soil Food Web. As mentioned previously, the SFW is extremely complex and redundant and is therefore referred to as a web rather than a simple chain. The SFW includes (from smallest to largest): bacteria, actinobacteria, fungi, protozoa, nematodes, enchytraeids, arthropods, and earthworms. This is not an all-inclusive list, but a sampling of the major players. More about each of these later in this chapter.

Bacteria are some of the smallest organisms in the soil and often the most numerous. Bacteria are fairly short-lived, opportunistic and serve as food for many other organisms. Because bacteria are so numerous (potentially billions per teaspoon of healthy soil) they have the capacity to have a rapid and significant impact on soil organic matter and plant nutrient availability. By changing the soil environment with tillage or inputs such as manure, fertilizer or plant residues, bacteria can react by multiplying quickly and taking advantage of changes in moisture, temperature, oxygen or food supply to quickly consume resources and alter the chemical and physical properties of the soil. Tilling soil when it is warm and moist, adds oxygen to the soil atmosphere while simultaneously fracturing soil aggregates, and consequently making soil organic matter available for bacterial consumption. Thus, organic matter in the soil is quickly "burned" by bacterial respiration and lost to the atmosphere as carbon dioxide.

The tragedy of tillage is that the organic matter that is most easily consumed by copiotrophic bacteria are the biologic glues that hold the soil into stable aggregates. It is in this way that tillage often results in bacteria burning the house down for the rest of the soil food web. After the tillage operation is complete, Nature steps in and expends tremendous energy to restore what has been destroyed. Energy that should be powering a smooth-running soil is instead used to rebuild it. It is because of this that large amounts of petroleum based energy in the form of fuel, fertilizer and pesticides are continuously applied to the soil of most agricultural systems to make up for the soil's reduced capacity to function.

Carbon Is Key

Much of modern agronomy is focused on feeding plants the nutrients they need to grow and produce a harvestable crop. As mentioned previously, the most limiting element in the soil for agricultural crop production is carbon. This should come as no surprise. Carbon is the building block for all life on Earth. Since soil is designed to function as a living system, carbon is essential to make it work.

There is plenty of nitrogen in the air, as 78% of the Earth's atmosphere is nitrogen, and there is plenty of phosphorous, potassium, and other elements in the organic and mineral fractions of the soil itself. It is not for lack of these nutrients in the soil that plants suffer nutrient deficiencies, but instead it is a lack of *available* nutrients and water. The majority of available plant nutrients are contained in, or made available by, the living fraction of soil organic matter. Up to 10% of what we refer to as soil organic matter is *living* microorganisms. It is this living fraction that fixes nitrogen from the air and decomposes everything from crop residues to dead animals to rocks, making nutrients available to plants. Carbon is a major and essential part of soil organic matter. *Quality* soil organic matter is what is lacking in most agricultural soils.

Soil organic matter often makes up less than 5% of the soil by weight, but controls *90% of soil functions essential for plant growth.* As soil organic matter increases from 1% by weight to 3% by weight, the water holding capacity of the soil doubles. 95% of the nitrogen and over 50% of the phosphorous in the soil is contained in soil organic matter. Between 19% and 54% of the cation exchange capacity (ability of the soil to hold onto plant nutrients) of a soil is due to soil organic matter (Hoorman & Islam 2010).

Soil organic matter includes everything from a living microbe to humus (the black part of soil formed from fully decayed plant material) and is critical to producing crops profitably. The majority of water and nutrient supplying capacity of the soil is based on the quantity and quality of soil organic matter. Our focus in agronomy should be on managing the soil as a biologic system, not managing it as a chemical system. The fundamental thinking that must change is from directly feeding the plants with fertilizer in a dysfunctional soil, to restoring a fully functioning soil that feeds the plants. We must learn to manage plants to feed the soil, so the soil *can* feed the plants. When fertilizers are added to the soil, they are first acquired by soil bacteria, who in turn supply the plants or other organisms with those nutrients. As mentioned in the introduction, managing for soil health is more about employing seeds to build the soil rather than using steel to disturb it.

The easiest source of food for soil microbes is the sugar exuded through the roots of living plants. The next easiest food source is dead plant roots. Followed by above-ground crop residues such as straw, chaff, husks, stalks, flowers, and leaves. When root exudates, dead roots, or plant residues are not available, soil microbes will feed on existing soil organic matter. When existing soil organic matter is the only source of food available for soil microorganisms, soil organic matter will decline in both quantity and quality.

Let us pause here and take a closer look at what is considered food for soil microorganisms.

Plant root exudates have been found to include amino acids, other organic acids and a wide array of simple carbohydrates (sugars) (Walker et. al. 2003). Exudates are substances plants secrete through their roots into the soil. These exudates can adjust the populations of soil microorganisms in the rhizosphere (area of soil immediately adjacent to a living root), help the plant deal with leaf harvest by grazing animals, change the chemistry (including pH) of the soil, and inhibit other plants from growing nearby. Plants control a great deal of the soil environment they are rooted in by sending chemical signals and supplying food to the organisms that live in the soil around their roots.

In most agricultural soils with severely diminished soil food webs, the exudates released by plant roots are not received very well by the intended soil organisms because many of them are not present. This causes the plant to expend additional energy on exudates in an attempt to feed a depleted SFW. Excessive soil disturbance simplifies the soil food web to include only a limited number of species; mostly bacteria. Thus, plants in a degraded soil often signal and supply food in vain for lack of the species of microorganisms that are supposed to be present in the soil. Plants will exude from their roots up to 70% of the carbon they fix from the atmosphere through photosynthesis. This is a very large investment on the part of the plant and exemplifies how plants are designed to work with their partners in the soil. Sadly, most of those soil partners are not present in quantity or diversity in most of our agricultural soils due to disturbance by tillage and excessive inputs. This results in great inefficiencies of crop production as crops call out to their absent partners in a nearly dead soil.

Excessive applications of nitrogen fertilizer only exacerbate this condition. As bacteria (Nitrosomonas and Nitrobacter, in particular) quickly acquire any fertilizer and convert it to plant available nitrates the plants exude few sugars from their roots as they do not immediately need the services of the rest of the soil food web. As the growing season progresses and the nitrogen from the fertilizer is gone from the soil (it either leached below the root zone, volated into the atmosphere or was taken up by the plants) the plants still require more nitrogen from soil organisms. This puts the plant in a difficult situation as it exudes sugars to feed the organisms that it had previously shunned. It takes time and energy for the plant to establish the necessary relationships with the desired soil organisms and thus crop yield is diminished at a time when it should be advanced. This scenario plays out every year across most of the cropland in the U.S. as farmers and agronomists attempt to feed the plants directly, not realizing how disruptive excessive fertilizer applications are to SFW-plant relationships.

Dead plant roots comprise the next most available source of food for soil microorganisms. This material consists of complex carbohydrates of cellulose and lignin. These materials are more difficult to digest by soil organisms than the easily assimilated simple sugars from living roots. By virtue of their position in the soil, dead roots are more accessible to soil microorganisms and often in a condition of higher moisture. But they nevertheless require more energy, enzymes, and nitrogen to break down into readily consumed simple sugars.

Other residues, left behind on the soil surface, are next in availability to soil organisms. Because of their position above the soil surface, they are less accessible and in a somewhat less hospitable place than residual roots *in* the soil. Plant residues on the soil surface are decomposed more rapidly by larger organisms, such as arthropods, that either consume the residues, or shred them into smaller pieces. The smaller pieces of residue provide a greater surface area and are more susceptible to digestion by fungi and other organisms. The digestive systems of most arthropods do not process plant residues completely and thus their droppings become viable food sources for fungi and other organisms. In this way, the soil acts as a multi-stage external rumen, or stomach, as it decomposes complex carbohydrates down to where nearly all the energy has been extracted from them.

The food of last resort for soil organisms is the soil organic matter it-self. Regrettably, in soils that experience significant repeated tillage, this is often the source of food most readily available to opportunistic soil bacteria. Crop production systems that rely on recurring tillage of the soil force the soil organisms to eat their own house, leaving them in a homeless state of dysfunction.

There is a growing acknowledgement by those in agriculture that the biology of the soil is important, if not critical, to crop production. However, not everyone is looking at soil biology in the same way. There are a number of biological products being developed in which particular organisms are be-ing isolated from the rhizospheres of plants based on how those organisms have proven to affect plant performance. While there is little doubt that some of these products will produce higher crop yields, producers must be careful not to fall onto a treadmill similar to that created by concentrated fertilizers. Leaving a soil in a degraded condition and supplementing any soil biological shortcomings with fertilizer or biological amendments will not necessarily set one on the path of restoring soil health. Without creating good soil habitat, and restoring the entire soil food web, your soil will not be-come self-sustaining. The addition of a particular organism from a biological amendment can either act as a tool to begin to restore soil health, or merely provide a crutch to help a disabled soil continue to limp along. Soil habitat is key. If you build it, the organisms will come. They just need an adequate home in which to become established and then thrive.

Below are some key points to consider regarding soil biology and soil organic matter:

- Every farmer has livestock. A handful of soil can contain 50 billion bacteria. There are at least 2,000 pounds of soil microbes in each acre of healthy soil; the weight of two cows. Soil microbes eat soil organic matter if no living roots are feeding them (Hoorman & Islam 2010). Biologic activity can warm the soil, similar to the many organisms respiring in a compost pile.

- Bacteria are only 20% - 30% efficient at keeping carbon in the soil. Fungi are 40% - 55% efficient at keeping carbon in the soil (Hoorman & Islam 2010). Fungi connect with plants and bring water and nutrients to them in exchange for sugar. Plant roots alone explore 1% of soil. Plant roots associated with fungi explore 20% of the soil (Hoorman & Islam 2010).

- Each pound of soil organic matter can hold 18-20 pounds of water (Hudson 1994). 1% organic matter measured in the soil can hold 1" of water.

- Soil compaction is a symptom of soil dysfunction that must be solved biologically. Cold, wet soils result from compaction. It is compacted layers of soil that do not allow excess water to drain downward in the soil. Compacted soil saturated with water is the reason soil warms slowly in the spring, not because residue on the soil surface is keeping the soil wet and shaded.

- Earthworm populations will suffer if soil is tilled and their food (crop residues) is buried and their cocoons are moved too deep or too shallow in the soil for them to hatch and grow into worms. Earthworms reproduce based on weight, not by age. The faster they grow, the earlier in their lives they are able to reproduce (Edwards & Bohlen 1977).

- Legumes (plants that bear their seeds in pods, i.e. peas and beans) cooperate with Rhizobia bacteria that capture (fix) nitrogen from the air. Because these types of plants can host bacteria that capture nitrogen from the air, they are typically rich in nitrogen and have a correspondingly low carbon-to-nitrogen ratio. Carbon to nitrogen ratio is a ratio of the mass of carbon to the mass of nitrogen in a substance. For example, a C:N ratio of 10:1 means there are ten units of carbon for each unit of nitrogen in the material. More about C:N ratios later.

- Grasses accumulate nutrients, including nitrogen, from the soil with their fibrous root systems and their C:N ratio is typically high when the plant is mature.

- Brassicas; which are plants of the mustard family, such as broccoli, cabbage, cauliflower, and turnip, have robust soil-penetrating root systems that open pathways deep into the soil.

- There is a great deal of collaboration that goes on between soil organisms and plants. In a legume-grass mixture of plants growing in the soil, legumes fix nitrogen which is often shared with grasses through fungi that connects the two plant types. Grasses accumulate phosphorous with fibrous roots and share phosphorous with legumes via the same fungal connection.

- Protozoa are predators (the coyotes of the soil) and eat bacteria (the mice of the soil) and excrete nitrogen as ammonium for plants to use. A 10% increase in soil temperature doubles microbial activity. Approximately 2% of soil organic matter will be mineralized (liberating the crop nutrients it contains) per year (Davidson & Janssens 2006).

- It is very difficult to build organic matter in the soil while performing repeated tillage. The physical disruption and aeration of the soil results in organic matter being consumed faster than it is produced.

Once again, the principal inhabitants of the soil food web are: bacteria, actinobacteria, fungi, protozoa, nematodes, enchytraeids, arthropods and earthworms. Bear in mind that bacteria, actinobacteria, fungi, protozoa and nematodes move about the soil in films of water within the pores of the soil. Without the right sized pores containing water, these smallest members of the SFW will become immobile and/or dormant.

Bacteria – As mentioned previously, bacteria are some of the smallest microscopic organisms in the soil and often the most abundant. They are short-lived, opportunistic and serve as food for protozoa, nematodes and others in the soil. Bacteria can multiply rapidly and impact the balance of plant nutrients in the soil quickly.

Bacteria can be free-living or establish symbiotic relationships with plants or fungi in the soil. Rhizobia bacteria living in the roots of legumes is the classic example of a bacteria–plant symbiosis. Rhizobia reside in the roots of leguminous plants and fix nitrogen from the air and share it with their host. The degree to which Rhizobia can fix nitrogen from the soil atmosphere depends, in part, on how efficiently air is exchanged in the soil. Stable soil aggregates and root and earthworm channels provide pores for gas exchange in the soil. Since the Earth's atmosphere is mostly nitrogen, the increased gas exchange through soil pores facilitates nitrogen fixing bacteria's access to this airborne source of nitrogen. A compacted soil, or soil with poor porosity resulting from soil aggregates being crushed by tillage, would limit the ability of a legume to fix nitrogen. This is not only because the leguminous plant might suffer growing in such conditions, but that the Rhizobia bacteria would not have access to as much nitrogen gas in the soil. Therefore, we cannot assume that a properly in-oculated legume will fix nitrogen to its full potential if soil porosity, and the corresponding exchange of gases with the above-ground atmosphere, is restricted.

There are also free-living nitrogen-fixing bacteria in the soil that may colonize the hyphae (threadlike "roots") of mycorrhizal fungi, feeding on the glomalin that the fungi make to coat their hyphae. Mycorrhizal fungi will of-ten pick up these organic forms of nitrogen that have been generated by bac-teria and deliver them to the host plant in exchange for sugar. By managing for improvement of soil health we can increase the capacity of soil organisms to fix atmospheric nitrogen and increase the amount of nitrogen available to our crops. This is another example of how a vibrant soil food web can feed plants rather than plants relying on us to provide supplemental sources of nitrogen.

Actinobacteria aka Actinomycetes – Actinobacteria are bacteria that grow as thin microscopic filaments in the soil, similar to fungi. They are also similar to fungi in that they have the capacity to decompose complex carbo-hydrates such as lignin and cellulose (i.e. corn stalks and wood). But because they are actually bacteria, and not fungi, they are more currently referred to as Actinobacteria. There are many species of these type of organisms in the soil. Some species such as the genus *Frankia*, can create symbiotic relation-ships with non-leguminous plants to fix nitrogen from the atmosphere, while other species of Actinobacteria are the source of some of our most valuable antibiotics.

Fungi - Mycorrhizal fungi live in the soil and connect with living plant roots to supply a great deal of water and nutrients to the roots of those plants. Their hyphae are barely visible to the naked eye, but can be viewed with a 10x hand lens. Mycorrhizal fungi are very sensitive to soil disturbance and rely on a living plant to sustain them. The living plant is key to support-ing soil fungi and building soil aggregates. Without living plants, vital soil aggregates will not be formed nor maintained.

Plants obviously benefit greatly from this arrangement, but mycorrhizal fungi don't enter into the partnership free of charge! Mycorrhizal fungi may consume up to 30% of the sugars a plant produces from photosynthesis during their association with the plant's roots. Flax and corn associate with mycorrhizal fungi and secure a fair amount of their phosphorous needs in this way. If an excessive amount of phosphorous fertilizer is added to the soil, it will often cause the association plants make with mycorrhizal fungi to be diminished. The hyphae of mycorrhizal fungi can explore ten times the volume of soil that plant root hairs can, and 100 times that of plant roots themselves. Mycorrhizal fungi can access water from very small pores in the soil and bring that water to plants during times when the soil is dry. The fungal hyphae are typically less than four microns in diameter, so they can get into very small spaces in the soil! Mycorrhizal fungi not only are capable of connecting the soil to the plant, but can connect to other fungi and other plants at the same time allowing nutrients to be shared between them all.

Plants in the mustard and lambsquarters families do not associate with mycorrhizal fungi much, if at all. When a crop like corn follows a crop like sugar beets that is non-mycorrhizal, the corn may suffer due to lack of mycorrhizal fungi populations recently active in the soil.

Excessive tillage can cause a major decline in mycorrhizal fungi populations because the fungi are very sensitive to soil disturbance that breaks up their hyphae. In undisturbed soil, mycorrhizal fungi will form extensive net-like webs in the soil. If the hyphae are broken by tillage, the fungi must begin again to grow new hyphae. This is a very energy intensive process for them. After fungi are disturbed, they commit a great deal of resources to restoring themselves instead of assisting their partners, the living plants.

Basidiomycetes are another type of fungi found in the soil that feed on dead plant material rather than connecting to living roots. They live in the rhizosphere, close by, but not directly connected to plant roots. Basidiomycetes, like other fungi, produce sticky sugary substances that glue soil particles together to form water stable aggregates. They also generate specialized enzymes that break down tough crop residues. These enzymes often perform an important step to break down some waxy crop residues like corn. Basidiomycetes and the presence of the sticky sugars they produce can be used in laboratory analysis as indicators of changes in soil aggregation. The health and vigor of basidiomycetes in the soil depends on a diversity of plants to provide a variety of residues that serve as sources of food for the fungi. It is equally important to not disturb the soil where basidiomycetes live. Basidiomycetes, like mycorrhizae, are very fragile and sensitive to physical soil disturbance.

No matter which type of soil fungi we are concerned with, the plant root associating mycorrhizal fungi, or the residue-decomposing basidiomycete's fungi, a diverse crop rotation and little or no soil disturbance are critical to their health and well-being. Crops living in a soil with healthy populations of mycorrhizae and basidiomycetes will benefit from the decomposition of old crop residues, improvement and maintenance of soil aggregates and the movement of nutrients and water through the many microscopic pipelines that both of these fungi create in the soil.

Protozoa – Protozoa are microscopic predators in the soil. They feed primarily on bacteria. Protozoa move about in water films in the soil to pursue their prey. Plant nutrients are cycled in the soil when microorganisms are born and when they die. Not many bacteria die of old age. They are often consumed by protozoa, nematodes, enchytraeids or earthworms in the soil. Without a diverse population of protozoa in the soil, nutrient cycling will be very limited. This is true in most agricultural soils that do not provide adequate habitat for the soil food web. This is why most current methods of agricultural production are reliant on the repeated application of fertilizer, compost or manure to sustain crop yields. Applying compost or manure to the soil does not by itself restore soil health without simultaneously maintaining favorable habitat for the soil food web. Adding compost or manure to the soil followed by tillage merely releases the nutrients contained in the added compost or manure while destroying soil aggregates. This will produce good crop yields in the short-term, but will not by itself restore soil health.

Nematodes – Nematodes are also predators in the soil. They feed on bacteria, fungi, protozoa, plants, and each other. Nematodes are microscopic, appear worm-like and move in a whip-like fashion in water in the soil. Some nematodes may switch from feeding on fungi to feeding on plants when fungi are not available. Other nematodes are specialized to feed on only one type of prey, such as bacteria. Nematodes, like protozoa, are important for nutrient cycling to occur in the soil. Without a variety of pores sizes and water films in the soil, nematodes will be very limited in their ability to move about and feed. As soil habitat is improved, the populations of various species of nematodes will increase and nutrient cycling will increase along with them.

Enchytraeids – Enchytraeids are analogous to tiny versions of earthworms and are sometimes referred to as pot worms. Enchytraeids are often white or light tan colored and are visible with the naked eye. They are predators of smaller organisms of the soil food web.

Arthropods – Arthropods have legs and include a wide variety of creatures, such as mites, springtails, beetles, spiders, and ants. They eat smaller organisms of the soil food web as well as plant material. They often serve a role as shredders, breaking plant material down into smaller pieces that are more easily accessed by fungi and other soil organisms. They also prey on most other inhabitants of the soil, in addition to plants and seeds. They can play a significant role in weed control by eating weed seeds present in and on the soil. Many arthropods live *above* the soil among plant residues. This layer of plant residues on the soil surface is sometimes referred to as the detritusphere (Beare, et al. 1995).

Earthworms – Earthworms eat most of the other members of the soil food web smaller than they are, create sizeable pores in the soil and create stable soil aggregates in the form of their castings (manure). The presence of earthworms is an easily observed positive indicator of soil health. Since earthworms reproduce based on weight, not age, the faster they grow the earlier they reproduce and increase in number. An earthworm's rate of growth is a function of its habitat. Undisturbed soil covered with plant residues with a plentiful soil food web will promote earthworm growth and reproduction.

A soil that provides beneficial habitat for all the members of the soil food web will allow them to build soil aggregates and restore the capacity of the soil to function. In particular, the capacity of the soil to infiltrate water and cycle nutrients is critical to the profitable production of plants.

CHAPTER 5

How Do I Restore the Health of My Soil?

P ut quite simply, restoring soil health requires a paradigm shift from thinking of the soil as dirt to thinking of soil as a living biologic system. Once this change of thinking occurs, it can be applied to any system of crop production. Most true paradigm shifts result in a reset to zero. In other words, you need to go back to square one. This can be a bit unsettling, as it was for me many years ago when I realized that, as an agronomist and soil scientist, I knew almost nothing about how healthy soil functioned. I understood the chemistry, physics and classification of soils, but not how it functioned. I was missing an entire dimension of the soil that was the most important to agricultural production and environmental quality. While this was a daunting revelation, I knew I needed to explore the concept of soil health further if I was going to actually help anyone solve their natural resource concerns.

I began writing and speaking about soil health in 1992. I recall my first soil health presentation that year, in Beulah, North Dakota. I began my exposition of how the soil truly functioned as a biological system and stated that "tillage is bad for the soil." My statement was immediately challenged, turning the presentation into a lively discussion. The soil paradigm shift I was proposing to the group that day was viewed by some as insanity, and to others as genius. Mostly the former.

As I persevered and learned all that I could about soil biology from farmers, ranchers, soil microbiologists, plant physiologists, mycologists, soil ecologists and others, I had my eyes opened to the truth about soil and how it functioned. In my campaign to teach others about soil health, I also realized that I was up against a 7,000 year old tradition of tillage and decades of addiction to the use of fertilizer. But the fact remained that nearly all of the history of agriculture consisted of the degradation of the soil in the pursuit of maximum crop yields. Something needed to change. Most in agriculture accept soil degradation as the cost of production and apply amendments and best management practices to prop up production and slow erosion. But I knew that treating the symptoms was not going to solve the problem of soil degradation and dysfunction. Only an understanding and appreciation for how the soil functions would save us from destroying the soil that feeds us.

To restore the health of your soil is to appreciate that the soil is habitat for the soil food web. If the soil is managed with an understanding of the habitat requirements for the SFW, the capacity of the soil to function can be restored. Regardless of the organism in question, habitat is simply a combination of food, water and shelter necessary for certain organisms to flourish. For example, in order for birds to flourish, they need food (typically insects and seeds), a source of water to drink and bathe, and cover to build their nests and rear their young. Habitat for the organisms of the SFW includes undisturbed soil that is always covered with plants and/or plant residues and is inhabited by a wide diversity of plants whose roots are alive as much of the time as possible. In effect, this means to manage soil as Nature would manage it. Wherever you are in the world, there is a native plant community that existed before humans made their mark on the land by tilling the soil. It is these uncompromised native plant communities that serve as the template we need to mimic in order to restore the soil to its full capacity to function.

Our role as soil managers should therefore be to make the soil the best habitat possible for soil microorganisms to thrive, build soil organic matter and feed plants. *This can be accomplished by disturbing the soil less, growing* _the greatest diversity of plants possible, keeping living roots in the soil as much as possible, and keeping the soil covered at all times._ This will provide the ideal habitat for soil organisms to thrive and associate with each other in order to build soil and restore water and nutrient cycles.

Keys to Restoring Soil Health

- Less Soil Disturbance
- More Plant Diversity
- Living Roots as Much as Possible
- Keep the Soil Covered at All Times

These four key principles to restoring soil health are applicable to any soil anywhere in the world. As you seek to develop your own approach to improving soil health, I encourage you to revisit these principles to see where your present approach has opportunities for improvement. By carefully examining your system of crop or livestock production against these four principles you should quickly be able to identify any weak links. You can then work to correct any shortcomings while monitoring your soil to see if its health is improving.

If your system of production relies on a significant degree of soil disturbance, you will need to mitigate for that disturbance by enhancing plant diversity, living roots in the soil and soil cover. If your system is lacking soil cover at some time, you may need to reexamine your crop rotation, add cover crops or mulch to correct the shortcoming. If you continue to deal with symptoms of weed, insect or disease pressure in your crops you may need to diversify the crops in your rotation or add a mixed-specie cover crop to your system. Regardless of your approach, by monitoring soil health indicators you can determine if your progress toward improving soil health is on track. _The proof will always be in the pudding._ Don't assume because you implement a no-till or organic crop production system that your soil health will improve. I have seen examples of many different systems (including no-till, conventional, and organic) succeed greatly, or fail miserably, at restoring soil health depending on how well the land manager understood the principles of soil health and how to apply them.

Let's examine each of these soil health principles in a bit more detail.

Less Soil Disturbance

There is evidence that humans have been tilling the soil for over 7,000 years (Lowdermilk 1953). In that time, there has not been a single tillage operation performed that directly benefited the soil. *Not one.* This is not to say that we should never perform tillage, but we should not fool ourselves into thinking it will directly benefit the soil in some way. Tillage immediately degrades the structure and function of the soil, period. Tillage is a tool, not a crop production system. Tillage is performed to manipulate the soil into a temporary condition to suit what humans wish to produce in the soil at the time, not to benefit the soil itself.

This is an extremely difficult concept for many of us to grasp. Tilling the soil is not a noble undertaking, it is destructive. In this regard, we must undergo a fundamental paradigm shift in our thinking of soil and agriculture in general. When the tool of tillage is employed to control weeds or manage crop residues, we must examine the damage it will cause and learn to mitigate for that damage if we are to restore the health of the soil. There are fifteen states in the U.S. that include an image of a plow in their state seal: Arkansas, Iowa, Kansas, Minnesota, Montana, Nevada, New Jersey, North Dakota, Oklahoma, Oregon, Pennsylvania, South Dakota, Tennessee, West Virginia, and my native state of Wisconsin. The plow is also front and center in the seal of the U.S. Department of Agriculture. The plow may have given us a start in building civilization, but it may ultimately be our undoing (Montgomery, 2007).

Soils on Earth formed and continued to develop for many years before humans learned to manipulate them. Until recently, humans have rarely had a positive effect on soils. Tillage of the soil disrupts and destroys soil habitat for the SFW. Tillage cannot be performed to benefit the soil in the name of loosening, amending, aerating, incorporating or otherwise modifying the soil. The surest way to restore soil health would be for us to walk away from it and never impose our will on it again. Nature would restore the soil and resume building it again as she always has. For perhaps the first time in human history, we now have the understanding to restore the soil rather than degrade it.

This leads us to the primary paradigm shift that must occur in the way we think about soil. Left to itself, the soil will heal and restore its capacity to function. We need to understand what the soil does and how it expends energy to restore itself. We need to understand what soil is striving to do, not what we are striving to do (produce crops) while using the soil as a medium for our desired means of production. By attempting to exercise control over the soil and the plants that grow in it, we almost always degrade the soil and waste a great deal of energy in the process.

My point by taking this perspective of the soil is not to dishonor anyone, including those that make a living from the land, *but to wake us up*. We need to understand how the soil was designed to function so we can use it as it was designed to function. As with anything, it is best to operate things the way they were designed to function, not only to get the most out of them, but for the continued capacity of that thing to function. The soil is no different. Understand how it was designed to function and operate it accordingly and the soil will maintain its capacity to function while producing our food and fiber.

Soil constructs and conducts itself biologically. This is accomplished through plants and soil organisms. Inasmuch as possible, we need to foster the ability of plants and soil organisms to help us grow our desired crops. This can be done by giving the soil what it needs. The soil does not need to be tilled. If we choose to use tillage as a tool to manipulate temporary soil conditions or plant species, then we need to understand how to mitigate for that disturbance so the soil is not resultantly degraded. Most current methods of agricultural production include unmitigated tillage resulting in continued soil degradation.

There is technology now available in the form of drills and planters that are designed to plant directly into undisturbed soil and plant residues. It is no longer necessary to disturb all the soil in a field simply because a drill or planter will not plant properly into anything but loose soil. At a small scale, such as my own no-till garden, I simply make a small slice with the blade-edge of a shovel or poke a small hole in the soil with a rod or stick just large enough to insert a seed or transplant. No other soil disturbance is necessary. Weeds are controlled with crop rotation, hand-weeding, mulch or between-row cover crops rather than tillage. In a larger scale scenario weeds can be controlled with herbicides (conventional or organic), residue management, cover crops, crop rotational diversity, or grazing by livestock.

More Plant Diversity

If a person went on a strict diet of eating nothing but donuts for a month, your health would most certainly decline. The same could be said of a strict diet of only broccoli for a month. A diet consisting of exclusively donuts, or exclusively broccoli, would most likely not result in good human health. Consuming a very diverse diet that includes many other foods in addition to donuts and broccoli, would be a much better approach to maintaining a healthy human. As a biological system, the soil is no different.

Like humans, the soil is a living biological system with a biome (a naturally occurring community of microorganisms living in harmony with a host) that requires a diverse diet to sustain a healthy condition. Plants feed the soil and the soil, in turn, feeds the plants. The diversity of plants grown in the soil determines the diversity of the diet received by the organisms that live in the soil. A diet supplied by only one or two plants in a cropping scheme is not enough diversity to sustain a healthy soil any more than a diet of solely donuts and broccoli could sustain us in a healthy condition. Cropping systems of continuous wheat or continuous corn, or even a two-crop system such as corn and soybeans, is not enough plant diversity to sustain a healthy soil (Bardgett & van der Putten 2014).

You only need to look at the native vegetation in the area around you to see what level of diversity Nature uses to keep the soil healthy. Many prairie or forest ecosystems include hundreds of different species of plants. While it may not be practical to include hundreds of species of plants in a cropping system, it would not be difficult to include tens of species of plants. Research regarding plant diversity (Tilman et. al. 2006) suggests that as few as five different species of plants can have a tremendous impact on total plant biomass production. Other experiments have shown that with increased plant diversity comes reduced weed pressure, improved plant nutrient cycling, and other improvements in soil function.

In 2006, the Burleigh County Soil Conservation District in North Dakota planted several cover crop plots of different species of plants as single species and as a mixture of six species growing together at a location just east of Bismarck. The plots were planted without tillage. Because of the drought occurring that year (1.8" of rain from April to the end of July), the single species plots were nearly dead by the end of July. However, the cover crop consisting of a mixture of millet, cowpea, sunflower, soybean, turnip, and oilseed radish was alive and green. This six specie mix had plants with six different rooting depths and structures, giving them access to moisture in different zones in the soil. The mixture fed and supported a greater diversity of organisms in the soil, most importantly fungi, which could bring water to the plants from places in the soil where roots could not reach. In the single species plots, the water was all being drawn from the same portion of the soil and was soon exhausted. The cover crop mixture that included a diversity of plants consisting of grasses, brassicas, and legumes had another advantage. The legumes in the mixture provided nitrogen to the whole community of plants and soil organisms and most likely transferred some of this nitrogen to the grasses and brassicas when all the plants were associated within the soil by the same mycorrhizal fungi. The cover crop mixture fed the soil a variety of root exudates in all regions of the soil for a longer period of time than any of the single-specie plantings. The oilseed radish plot produced 1200#/ac. of above-ground biomass, Purple Top Turnip 1400#/ac., Pasja Turnip 2000#/ac., Soybean 1300#/ac., Cowpea 1800#/ac., Lupin 1100#/ac. and the mixture of six species 4600#/ac. The mixture of six species together out-produced any of the single species planted alone by more than double! This is but one example of the power of plant diversity.

Work in South Dakota (Beck 2005), in addition to that by Tilman, has shown that plant functional groups (cool-season grass, warm-season grass, cool-season broadleaf, warm-season broadleaf and legumes) are also important to diversity from a standpoint of soil health and agronomic cost of production. In short, give Nature the diversity she desires and you will be rewarded with increased crop production and profitability.

There is a crop rotation diversity index, developed by Dr. Dwayne Beck of Dakota Lakes Research Farm (near Pierre, SD) that can help in assessing your crop rotation plant diversity. The index and method of calculation are described in detail at Dakota Lakes Research Farms website http://www.dakotalakes.com/. The index can be useful to compare different rotations to see which has a greater index value. Generally speaking, a crop rotation diversity index of 2 is considered a minimum for a rotation that will help limit weed, insect, and disease symptoms and contribute toward improved soil health.

Living Roots as Much as Possible

The most easily obtainable source of food for soil microbes is the sugar exuded through the roots of living plants. This is the reason why maintaining living plant roots in the soil for as much time as possible is important. If left to its own devices, Nature will cover the soil with a variety of plants whenever plant survival is possible. This is to assure that the soil is fed adequately to be able to maintain itself and support plants in the future.

Some traditional crop production systems include fallow periods where no plants are allowed to grow in the soil for a certain amount of time. While brief periods with no living plants may be advantageous to break weed, insect or disease cycles or accumulate additional soil moisture for the next crop, these periods are not directly advantageous to the soil and thus call for some degree of mitigation. Extended periods of time in keeping the soil plant-free (aka summer-fallow in the Great Plains) starve the soil organisms of food and typically leave the soil exposed without adequate cover. Fallow Syndrome is a well-documented phenomenon where crops that would typically benefit from association with mycorrhizal fungi suffer yield loss after a season of fallow (Wetterauer & Killorn 2013). This is due to the lack of living roots that would support the fungi and thus the fungi decline and go dormant. This means that the crop that trails fallow in the rotation will not have an adequate population of mycorrhizal fungi to associate with and to supply it with water or phosphorous. The result is a poor performance by that deprived crop.

A great deal of erosion by wind and by excessive microbial decomposition of organic matter (aerobic erosion) has resulted in soil organic matter levels in prairie soils declining to less than half of their original readings when first converted from sod to agricultural production. In this way, the soil organic matter was literally mined of its nitrogen and phosphorous for crop production purposes. As organic matter rapidly decomposes under tillage, the carbon dioxide leaves the soil as a gas, but the nitrogen once contained in the organic matter remains in the soil. As such, tilled season-long fallow did not conserve moisture as much as it liberated nitrogen from the native soil organic matter. This is one of many examples of how agriculture has been extractive of the soil.

In the history of agriculture in the U.S. there are few persons that can be regarded as stewards of the soil, not necessarily out of negligence or greed, but out of ignorance of how the soil was designed to function. Now that we know better, we need to do better and restore what we have taken from the soil.

Living plants are the only practical and economical way to restore the soil (Savory 1988). This is because the energy that powers living soil comes from the sun through the pathway of photosynthesis. Any other attempt to restore soil organic matter, biology, structure, and function through the application of soil amendments such as fertilizer, manure, compost or other carbon source would be prohibitively expensive or impractical except on a small scale. Such amendments could be valuable tools to assist in restoring soil health, if applied as a part of a larger strategy based on practicing the principles of soil health.

With the understanding of the critical role living plants play in soil restoration we can look at crop rotations, cover crops and polycultures in a new light. We can find opportunities to include a greater diversity of plants to grow in the soil for a greater amount of time within the environmental conditions of a given location. We can replace expensive inputs with actions that improve the soil and reduce the symptoms that we have created from degrading the soil.

Keep the Soil Covered at All Times

I don't like to see the soil. This may seem like an odd statement for a soil scientist to make, but it's true. If improving soil health is your goal, you should not see the soil very often. Soil should always be covered by growing plants and/or their residues, and therefore, rarely be visible from above. This is true regardless of land use (crop, hay, pasture, forest, range or garden).

As mentioned earlier, managing for soil health is mostly a matter of maintaining suitable habitat for the myriad of creatures that comprise the soil food web. Regardless of the crop production system, soil cover cannot be taken for granted. Because of harvest methods, amounts of residue produced and carbon:nitrogen ratios of various crop residues, there are times when soil cover may be lacking. Carbon:nitrogen ratios will be discussed in more detail later in this chapter.

Why is soil cover so important for improving soil health? Soil cover con-serves moisture, intercepts raindrop impacts, suppresses weed growth, and provides habitat for members of the soil food web who spend at least some of their time *above*, not *in*, the soil.

It has been my personal observation that in most situations, more than 65% of the soil must remain covered to limit evaporation of water from the soil surface. Bare and relatively dry soil heats up quickly in direct sunlight, and the hotter the soil gets, the faster water will evaporate from it. This not only wastes water that could be used for crop production in drier climates, but evaporating water leaves salts behind at the soil surface making the soil less hospitable to young seedlings. If the soil gets too hot, all but the most heat-loving organisms will go dormant, hindering many vital soil processes. Residue cover also limits the drying effect from wind and traps snow during the winter in places where snowfall occurs. Conditions where soil and plant residues retain moisture also fosters the colonization of fungi on the plant residues that begin to break down com-plex carbohydrates, making those residues easier for other organisms to use as food.

Soil cover consisting of plants or plant residues also protects soil ag-gregates from receiving a physical beating by raindrops. A healthy soil with water-stable aggregates held together by biologic glues may be able to with-stand wetting by the rain, but may not be able to withstand an extended pounding from the impact of raindrops. If water-stable soil aggregates are going to allow water to infiltrate into the soil they have to be protected from the elements (wind and rain in particular) with living plants or plant residues. A mulch of plant residues on the soil also suppresses weeds at critical times in the growing season to allow the desired crop to gain an advantage. Mulch may be applied to the soil surface in a number of ways. It may be applied manually using hay or straw, derived by laying a cover crop down with a roller or the hooves of livestock, or spreading crop residues with the same piece of equipment used during the harvesting operation, such as a straw and chaff spreader on a combine.

Plant residues on the soil surface also provide habitat for many species of arthropods that often begin the process of residue decomposition by shredding residues into smaller pieces (Nardi 2003). Not all members of the soil food web are microscopic. If these arthropod "shredders" have good residue as habitat, they can increase residue decomposition (and therefore nutrient cycling) up to 25% (Culliney 2013). The rate of residue decomposition can be managed quite handily without tillage, but instead by soil biology with suitable habitat. Keeping the soil covered, while allowing crop residues to decompose so their nutrients can be cycled back into the soil, can be a bit of a balancing act. Producers must give careful consideration to their crop rotation (including any cover crops) and methods of residue management if they are to keep the soil covered and fed at the same time.

If crops that produce low amounts of residue are grown to diversify crop rotations, or increase profitability, producers need to keep in mind how important it is to provide cover for the surface of the soil by conserving residue from other crops in the rotation. Crop residues covering the soil surface help to reduce disintegration of soil aggregates. It is also important to moderate soil temperatures to maintain favorable habitat for much of the microscopic life that lives and works in the soil. Producers need to weigh the value of crop residues that may be removed by grazing, baling or other methods against the value of those residues remaining on the soil surface. If excessive amounts of crop residue covering the soil surface is a concern when planting into cold soil in the spring, planting equipment may be fitted with residue managers that gently sweep residue away from a narrow strip where the seed will be planted. This will expose only the seeded strip of soil so it can be warmed by the sun, and leave the rest of the soil undisturbed and covered. As the years pass, soil managed under a system that maintains soil cover with a diverse crop rotation becomes more biologically active, cycling nutrients, building organic matter and even warming the soil through increased respiration, similar to the heat generated in a compost pile. This increased biologic activity will also cause crop residues to decompose more rapidly, which will need to be taken into account in order to retain adequate crop residue on the soil surface. Higher rates of microbial respiration are a good thing if the respiration is coming from organisms building soil organic matter rather than organisms opportunistically devouring soil organic matter after a tillage event.

Developing Your Own Approach to Restoring Soil Health

As mentioned previously, restoring soil health requires that you change your perception of the soil from dirt, to thinking of the soil as a living biologic system. Once this change of thinking occurs, it can be applied to any system of crop and/or livestock production. We need to think in terms of, "What would Nature do?" To see what Nature is already doing, find a native plant community in your area and see what plants are present and what the soil looks like. This can serve as your template to make comparisons with your current system of production.

Instead of looking at your system to maximize yield of a particular crop, look at making the soil the best habitat possible for soil microorganisms to thrive and build soil organic matter and feed the plants. Carefully examine your system of production and analyze it according to the four principles of managing for soil health. During your examination you need to borrow the most-asked question of three year-olds, "Why?" This self-examination of your agricultural system and any resulting shift in thinking may not be easy. The most difficult process you will go through to restore soil health on your land will be in your own mind.

Less Soil Disturbance – What types of tillage are you currently performing and why? What could you substitute for some or all of the tillage? What past actions created the symptoms you are attempting to control with tillage today? By understanding that each tillage operation is destructive to the soil, you must seriously and creatively reduce tillage as much as possible. Once you realize that tillage does not solve any problems you have with your soil, but only temporarily treats symptoms that prior tillage has likely created, you will be able to climb out of the box that you have placed yourself in, and look at the soil and your production system as a whole. If you are performing any tillage operation with the thinking that you are improving the soil, you are wrong.

Be sure to also take a look at what chemical disturbances your system may inflict on the soil. The use of certain pesticides and fertilizers may have little effect, or a significant effect, on the life in the soil. This can depend on the type, rate, time and frequency that a particular product is used. Question each product and do a little research on how each one might affect non-target species living in the soil. Something as simple as a heavy application of fertilizer can not only injure soil organisms directly, but disrupt the relationships those organisms may have with the plants they associate with, causing harm indirectly. Much remains to be learned about how pesticides and fertilizer affect soil organisms. Hopefully, researchers will be able to answer more of these questions in the near future.

More Plant Diversity – This is the most underestimated and least practiced aspect of restoring soil health. Most farms have become specialized to produce only one or two crops and almost always as monocultures (one specie of plant grown on a field at a time). This is not how Nature produces plants or how soils were designed to function. You will notice in your local example of a natural native plant community that hundreds of species are all growing in the same locale with different species growing right next to each other. It is little wonder that Nature is continually doing her best to increase the species diversity in the fields of our agricultural systems! She does this with what we refer to as weeds. Weeds are Nature's response to injury or imbalance in the soil. Nature strives to produce the most from the resources available and maintain a viable and resilient soil ecosystem. When we till the soil and plant only one species of crop, we are wounding Nature and conducting a hostile takeover. I do not use this analogy to appeal to your emotions, but to appeal to your intellect and your finances! No one has deep enough pockets to win a war against a natural system that has been established long before humans learned to dig in the soil. Many farmers have gone bankrupt trying to fight this natural order and I suspect many more will follow. My purpose in writing this book is to help keep you from becoming one of them.

Increasing plant diversity on your farm or ranch will help reduce the pressure Nature is exerting to grow what we call weeds. Increasing plant diversity will simultaneously increase the diversity of the organisms in the soil by providing them with a more diverse diet. This will improve aggregate stability, water infiltration, and crop nutrient cycling.

To assess the diversity of your current crop rotation and help you develop a more diverse rotation, I again encourage you to check out the crop rotation diversity index, developed by Dr. Dwayne Beck of Dakota Lakes Research Farm. The index and method of calculation are free and described in detail at Dakota Lakes Research Farms website http://www.dakotalakes.com/.

Again, a crop rotation diversity index of 2 is considered a minimum for a rotation that will help limit weed, insect, and disease symptoms and contribute toward improved soil health. This is true in farm fields as well as gardens.

Another excellent approach to increasing plant diversity on your farm or ranch is to plant cover crops. Cover crops are defined as a crop grown for the protection and enrichment of the soil. There are many purposes that cover crops can serve, from biological nitrogen fixation and pest control to improving soil structure and attracting beneficial insects. A particular cover crop can serve multiple purposes simultaneously if thoughtfully designed to do so. For most purposes, multi-specie cover crops are much more effective than single-specie plantings.

Green Cover Seeds, owned and operated by Keith and Brian Berns of Bladen, NE has some excellent resources for developing multi-species cover crop mixtures https://greencoverseed.com/. Most local Soil Conservation Districts can also provide technical assistance in developing cover crop seed plans. As mentioned previously, as few as five different species of plants can have a tremendous impact on total plant biomass production and soil health. When developing multi-specie cover crop mixes, consider at least five species of plants in the seed blend for a given planting.

No discussion of plant diversity would be complete without mentioning polyculture. Polyculture is growing more than one specie of plant at the same time in the same place. I have personally witnessed field pea and mustard grown together and harvested successfully in an economic manner. There are many other possible combinations of plants that can benefit the soil by being grown together for grain or forage. An internet search for "polyculture" can provide sources of additional information on polyculture food and fiber production at various scales.

Living Roots as Much as Possible – As pointed out earlier, the easiest source of food for soil microbes is the sugar exuded through the roots of living plants. Maintaining living plant roots in the soil for as much time as possible keeps the soil food web fed and active, building soil and cycling plant nutrients.

We must remember that every farm has livestock; trillions of head of them in every acre. Every acre of healthy soil has the equivalent weight of two or more cow's worth of microorganisms living in the soil. And like all livestock, they need to be fed. This is why it is vital that each field host a diversity of plants that occupies the soil with living roots and covers it from above. A diverse crop rotation, particularly one including cover crops, can provide the soil with the food and cover it needs to feed and care for the "two cows" that live within each acre of soil. Leaving a field bare or fallowed means the underground herd of microbes will not be fed and the soil functions dependent on those organisms will decline accordingly. The productivity and profitability of your farm or ranch depends how well your livestock are fed above and below ground!

Living roots means living plants that are conducting photosynthesis and capturing energy from the sun and moving it into the soil. Root exudates have been described as "liquid carbon" (Jones 2008). Think of it as filling the feed bunk for the underground herd. Without this feed the underground herd will suffer and the soil along with it. Much agricultural land has lost its capacity to function because the soil is not fed an adequate diet from living plants.

Keep the Soil Covered at All Times – We need a roof on the soil "house" to provide suitable habitat for the soil food web. Soil cover conserves moisture, deflects wind, intercepts raindrops, suppresses weed growth and provides habitat for members of the soil food web who spend at least some of their time above the soil. It is difficult to restore soil health and build soil aggregates in an unprotected soil. Ray Archuleta describes bare soil as "naked, hungry, thirsty and running a fever". Don't allow your soil to be any of those things.

Keeping the soil covered is a process of managing plants and their residues. Plant residues need to be left behind on the soil to provide cover and then decompose to allow nutrients to cycle, while not leaving the soil uncovered before the next set of plants can grow and cover the soil again. There are many ways to accomplish this, from hand-placed mulch to no-till farming or managed grazing. One of the keys to simultaneously managing soil cover and residue decomposition is an understanding of carbon to nitrogen ratios (C:N) of plants and their residues.

The C:N ratio of everything in and on the soil can have a significant effect on crop residue decomposition. This is why it is important to understand these ratios when planning crop rotations or including cover crops in a system of production. There may be times you would like residue to remain on the soil and times you would like it to disappear. This can be accomplished without tillage by managing C:N ratios.

Soil microorganisms have a C:N ratio near 8:1. They must acquire enough carbon and nitrogen from their environment to maintain that ratio of carbon and nitrogen in their bodies. Because soil microorganisms' burn carbon as a source of energy, not all of the carbon a soil microorganism eats remains in its body, a certain amount is lost as carbon dioxide from respiration. To acquire the carbon and nitrogen a soil microorganism needs to stay alive (body maintenance + energy) it needs a diet with a C:N ratio near 24:1. 16 parts of carbon are used for energy and 8 parts for maintenance. It is this C:N ratio of 24:1 that rules the soil! See the appendix for a listing of C:N ratios of some common plant residues.

If a microbial foodstuff such as mature alfalfa hay (C:N ratio of 25:1) is added to the soil, the soil microorganisms will consume it relatively quickly with essentially no excess carbon or nitrogen left over. The hay has an almost perfect balance of carbon to nitrogen (24:1) that soil microorganisms need.

What would happen if we added a foodstuff with a higher C:N ratio to the soil, such as wheat straw with a C:N ratio of 80:1? Since wheat straw contains a greater proportion of carbon to nitrogen than the 24:1 balanced diet soil microorganisms require, the microbes will have to find additional nitrogen to go with the excess carbon to be able to consume the straw. This additional nitrogen will have to come from another source of nitrogen in the soil. As soil microorganisms acquire this additional nitrogen, it is tied-up in their bodies. This situation could create a deficit of nitrogen in the soil until some of the organisms die, decompose, and release nitrogen, or some other source of nitrogen becomes available in the soil.

Conversely, let's look at what would happen if we added a foodstuff with a lower C:N ratio, such as a fresh hairy vetch cover crop with a C:N ratio of 11:1. Since the vetch contains a lesser proportion of carbon to nitrogen than the 24:1 diet soil microorganisms need, the microbes will consume the vetch and leave the excess nitrogen in the soil. This surplus nitrogen in the soil will be available for growing plants, or for soil microorganisms to use to decompose other residues that might have a C:N ratio greater than 24:1. This could result in bare soil if enough nitrogen is present to decompose the existing residues present at that time.

Everything else being equal, materials added to the soil with a C:N ratio greater than 24:1 will result in a temporary nitrogen deficit, and those with a C:N ratio less than 24:1 will result in a temporary nitrogen surplus. This is why composting operations strive to achieve a blend of materials with a C:N ratio of about 30:1, so the resident microbes can readily decompose the compost pile leaving a little food and structure left over to feed and shelter the microbes after the compost is applied to the soil (Lavelle & Spain 2005).

Next, let's examine C:N ratios from a practical perspective for crop production and building soil health.

C:N Ratio Effects on Soil Cover

The faster crop residues are consumed by soil microorganisms the less time those residues will be covering the soil surface. While it is important to maintain soil cover, it is also essential that those same residues decompose to release plant nutrients. It is important to understand crop residue C:N ratios to maintain soil cover when desired, yet allow the cover to ultimately break down and the nutrients be recycled.

A cropping system of continuous wheat certainly can provide good soil cover, as wheat produces a fair amount of residue with a relatively high C:N ratio (80:1). However, such residue will decompose relatively slowly. By adding a lower C:N ratio crop such as hairy vetch (11:1) to the rotation, a surplus of nitrogen will become available to the soil microorganisms, allowing them to break down the wheat straw more quickly.

C:N Ratio Effects on Nutrient Cycling

It should now be apparent from the discussion of C:N ratios and soil cover that management choices must strike a balance between crop residues covering the soil and residue decomposition for nutrient cycling. Managing residues so they cover the soil when a growing crop is not providing soil protection requires some planning and experimentation to achieve a proper balance. If crops with high C:N ratios are grown too frequently in the rotation, residues will accumulate on the soil surface. Nitrogen for crop growth may be restricted unless supplemented with other sources of nitrogen. This may result in poor crop performance during times when soil microorganisms monopolize soil nitrogen while working to decompose high C:N ratio crop residues.

Influence of Cover Crops on C:N Ratios

Cover crops included in a crop rotation can help manage nitrogen and crop residue cover on the soil. A low C:N ratio cover crop containing legumes (pea, lentil, cowpea, soybean, sunn hemp, or clovers) and/or brassicas (turnip, radish, canola, rape, or mustard) can follow a high C:N ratio crop, such as corn or wheat, to help the corn and wheat residues decompose, allowing nutrients to become available to the next crop. Similarly, a high C:N ratio cover crop that might include corn, sorghum, sunflower, or millet can provide soil cover after a low residue, low C:N ratio crop, such as pea or soybean.

Understanding carbon to nitrogen ratios of crop residues and other material applied to the soil is important to manage soil cover and crop nutrient cycling. A farmer I worked with some years ago in western North Dakota was interested in planting a multi-species cover crop to provide some fall grazing for his cattle and to improve soil health. He planted the cover crop, without tillage, following the harvest of winter wheat. The cover crop mix included turnip, a plant with a C:N ratio significantly below 24:1. Due to an error in communication regarding the mix, a planned planting rate of .7 pounds of turnip seed per acre was interpreted as 7 pounds of turnip seed per acre. The population of turnips across the field was tremendous. Due to the resulting low C:N ratio for the vegetation that remained after a moderate degree of grazing and a persistent snow cover during the winter, nearly all of the residue that was present in the fall had disappeared by spring, leaving the soil nearly bare. Using the low C:N turnip residue as food, the organisms in the soil consumed nearly all of the old wheat residue. This taught us that residue can easily be removed by soil biology if we are not cognizant of the C:N in a given situation. One could also apply this concept if it became necessary to reduce soil cover without using tillage to do so.

The four key principles of restoring soil health are applicable to any soil in any climate. As you seek to develop your own approach to improving soil health, I encourage you to revisit these principles to see where your approach might have room for improvement. By carefully examining your system of crop or livestock production against these four principles you should quickly be able to identify any weak links. Don't experiment on your whole farm or garden. Pick a portion you are willing to risk and test new methods on that area first. You can then assess how well those methods perform before applying them to additional areas.

As mentioned previously, if your system of production includes a significant degree of soil disturbance, you should seek to mitigate for that disturbance by enhancing plant diversity, adding living roots in the soil, or providing more cover on the soil. If your system is lacking soil cover at some time, you may need to reexamine your crop rotation, add cover crops or mulch to correct the shortcoming. If you continue to deal with symptoms of weed, insect or disease pressure in your crops you may need to diversify the crops in your rotation or add a mixed-specie cover crop. Regardless of your approach, by monitoring soil health indicators you can determine if your progress toward improving soil health is on track. Keep an open mind and reexamine the principles of restoring soil health until you develop an approach that works for you on your farm.

Additional Concepts for Restoring Soil Health on Cropland

Cropland is considered to be land where annual crops are planted each year. If tillage is part of a production system, cropland is typically where the most tillage occurs. Most tillage practices on cropland are performed to loosen the soil so conventional planting equipment will be able to place seeds or transplants properly in the soil and to control weeds. With the advent of no-till drills and planters, disturbing the soil to prepare a seedbed is no longer necessary.

Tillage is often performed to kill weeds that have begun to grow and therefore take the place of a desired crop. Tillage for weed control not only degrades the soil, it replants more weed seeds. After a few years of no tillage and controlling weeds with herbicides, flame-weeders, grazing livestock, manual weed pulling or suppression with mulch, weed pressure from annual plants declines to ever decreasing levels. This process of weed decline usually takes three to five years to accomplish (Hatfield 2016). It is during this same time that the biology of the soil changes to restore stable soil aggregates and improve plant nutrient cycles. I have observed this firsthand as part of the Southwest North Dakota Soil Health Demonstration Project in Dunn County (Eisenbraun 2011). One must be willing and able to take a leap of faith to stick with an approach that follows the principles of restoring soil health to give the soil time to show signs of recovery. We have been degrading the soil for many years in the United States. We must now be willing to take three to five years to allow it to turn in a positive direction toward renewed health. Be patient and have faith that Nature will restore your soil. You just need to give her the time and opportunity to do so.

For a crop rotation to have a chance to restore soil health, it must have at least three different crop types in sequence on a given field. The four basic crop types include: cool season grass, cool season broadleaf, warm season grass and warm season broadleaf. The greater diversity of crop types you have in rotation or crop types in a cover crop the faster the soil will respond in a positive way. This is often referred to as speeding up biologic time. Crop rotation diversity can move the process of restoring soil health along at a faster pace than one could with a crop rotation of only a few different species.

Another approach to increase the rate at which soil health is improved on cropland is to include grazing by livestock. One technique is to plant a multi-specie cover crop for the purpose of grazing. The multi-specie cover crop increases plant diversity in a crop rotation, extends the time that the soil hosts living plants and helps provide cover for the soil. By allowing livestock to consume half of the above-ground cover crop forage and trample much of the remainder onto the soil surface, the soil biology will be fed magnificently. When the plants are grazed, they produce additional root exudates to feed the soil food web as the plants strive to recover from being grazed. The plant residues trampled onto the soil surface, along with the manure and urine from the livestock, feed and cover the soil in a very favorable way. However, if livestock are allowed to remain in the field too long, the soil may become compacted and left with inadequate cover. The animals must be restricted to a small enough area at a time to achieve moderate grazing of the plants and trampling of the residues. Then, the animals need to be moved on to the next area so overgrazing or excessive trampling does not occur. Careful management of the livestock is critical to making this technique successful, but the soil health benefits can be tremendous.

Additional Concepts for Restoring Soil Health on Hay Land

Hay land is considered to be land where perennial plants are grown for the purpose of mechanical harvest for hay. Land that is in hay production for many years without the return of manure or plant residues suffers because the soil lacks food, cover and plant diversity. The forage that is produced is repeatedly exported off of the field and typically not returned. Land in hay production needs a diversity of plant species and residue to cover the soil.

An alternative to harvesting a field for hay each year is to alternate between properly grazing and haying from year to year. This leaves more plant residues on the soil surface and cycles more nutrients and biologic food back to the soil. Another effective strategy is to feed the hay on the field it was harvested from so that the residues and nutrients are cycled immediately back where they came from. This technique is often referred to as bale grazing (Manitoba Agriculture, Food and Rural Initiatives 2008). Bale grazing involves placing bales out in the field from where the hay was harvested and allowing livestock to consume the bales there, rather than in a feedlot. Any hay that is not consumed is returned to the soil surface to feed the underground herd of soil organisms. The manure and urine from the above-ground livestock is all left on the field as well.

Additional Concepts for Restoring Soil Health on Pastureland

Pastureland is considered to be land where introduced (non-native) perennials are grown for harvest by grazing livestock. One might think that soil health would be ideal in such a situation where there is no tillage, and living roots are present the majority of the time. However, soil health is just as easily degraded on poorly managed pastureland as on cropland. Pastureland often suffers from a lack of plant diversity as it is typically managed for one or two species of plants. Pastureland may also have bare soil between plants due to overgrazing, often resulting in poor plant canopy and little residue on the soil surface. Increasing the diversity of species on pastureland by inter-seeding other perennials or occasionally planting a multi-specie cover crop (without tillage) will add diversity to the soil to build soil health.

The key to managing pastureland for plant and soil health is to manage how plants are grazed and allowed to recover. Repeated grazing of the plants without time for them to recover results in diminished root systems and soil collapse. This can lead to compaction of the soil and water runoff from the paddock. Lack of grazing results in plants that are not stimulated by being bitten by livestock. The soil food web is thus not sufficiently fed by root exudates and the effective surface area of plant leaves conducting photosynthesis is reduced. All plants must be grazed and then allowed to recover before being grazed again. The amount of time between grazing events is a function of the species of plants involved, temperature and precipitation. The general principles that apply on rangeland also apply to pastureland. The Grazing Handbook, mentioned below under the rangeland topic, describes the ecology of grazing plants and how to manage them for optimum plant health and productivity, and for restoring and maintaining soil health.

Additional Concepts for Restoring Soil Health on Rangeland

Rangeland is considered to be land where native plant communities are harvested by grazing livestock. As with pastureland, rangeland often suffers from over-grazed plants or under-grazed plants, often side-by-side, or at least in the same paddock. The Grazing Handbook, available on the North Dakota State University Dickinson Research Extension Center website at: https://www.ndsu.edu/agriculture/ag-hub/research-extension-centers-recs/dickinson-rec/research/rangeland describes the ecology of grazing plant communities and how they can be restored and maintained with a corresponding positive effect on soil health.

Since rangeland consists primarily of perennial plants (plants that live for multiple years), there are living roots in the soil much of the time. However, those plants need to be grazed to stimulate them to exude sugars to feed the soil food web. After grazing, those plants then must be allowed to recover before being grazed again. Allowing livestock to graze plants at will does not allow time for plant recovery to occur. Plants in the paddock that are bitten will begin to recover and be green and tender, attractive for the livestock to bite them again too soon. The plants that are repeatedly bitten without time to recover decline. Plants in the same paddock that are not bitten become mature, less palatable and less attractive for the livestock to graze. The ungrazed plants begin to suffer for lack of stimulation and subsequently decline. Without proper management of the livestock to assure all plants are bitten and all plants recover, all of the plants will decline and the health of the soil with it.

The key to proper grazing management is not to allow livestock to graze plants until the plants are ready, either after breaking dormancy and growing to a certain stage in the spring, or since the last time the plants were grazed during the growing season. The livestock need to be concentrated into tight enough groups to encourage them to compete with each other and bite all of the plants. They then need to be moved to a new area and not allowed to return to the bitten plants until after the plants have adequately recovered. Most traditional grazing management allows livestock access to the whole paddock for a whole season and thus plants are not properly grazed or allowed to recover. The area to be grazed needs to be allocated to the livestock with fences, herding, water or other means to achieve the proper stimulation and recovery of the plants. In this way the plants will thrive and feed the soil, keeping both the plants and soil healthy. The concept is a simple one, but implementation requires thoughtful planning and execution.

Additional Concepts for Restoring Soil Health on Forestland

Forestland is considered to be land where trees are grown for lumber, paper pulp, or other forest products. Forest managed to maintain the greatest number of plant species and incurring the least amount of physical soil disturbance will help restore and maintain soil health. Your local conservation district can provide assistance or direct you to other technical resources to learn about properly managing forestland to restore and maintain soil health.

Additional Concepts for Restoring Soil Health in Your Yard, Garden, Vineyard, or Orchard

I have not tilled my western North Dakota garden, nor added any synthetic fertilizer, for over 15 years. I grow a wide variety of crops, from apples to zucchini with great success. Seeds are planted in tiny furrows made by slicing the soil open with the edge of a shovel blade, or dropped into holes poked in the soil with the end of a metal rod. Weeds are pulled by hand and then the soil is mulched with fresh grass clippings and, later in the summer, mulched with straw to maintain 100% soil cover. Irrigation water is added when needed.

The best place to begin a garden without tillage is by killing an area of existing sod (lawn). This can be accomplished with herbicide, covering the area with landscape fabric, a tarp, old carpeting or heavy cardboard for several weeks to kill the existing plants. Once the plants are dead, seeds can be planted and mulch applied between the planted crops as mentioned above. In this way, you will not be battling annual weeds, which would typically be the case in a previously tilled garden. Part of my garden was historically tilled soil and part of it reclaimed from grass sod. All of it is currently productive with minimal weed pressure. Undesirable plants are pulled from the soil and composted or added to the mulch between crops. The soil is not tilled (even using a hoe is tillage) to control weeds. Such soil disturbance will only plant more weed seeds and disrupt soil aggregates. The most difficult part of gardening without tilling the soil is to realize that the soil will become softer the longer it is left undisturbed and kept covered with mulch. Seeds or transplants will grow just fine when placed in a small scratch or small hole poked or dug in the soil.

Regardless of the land use, the principles of restoring soil health are the same:

- Less Soil Disturbance
- More Plant Diversity
- Living Roots as Much as Possible
- Keep the Soil Covered at All Times

CHAPTER 6

Problems, Symptoms, Goals and Tools

I n my years of teaching folks about soil health, I have often observed how individuals lose track of the goal of restoring soil health in their fascination with tools or a fixation on the symptoms of dysfunctional soil. Arguments often develop over which tools or systems are best and/or how to use them. Tools are useless except in the hands of a skilled worker with a blueprint of what is to be constructed. Everyone from producers to policymakers continue to struggle to understand the difference between goals, tools, symptoms, and problems when it comes to natural resource management. There is no right or wrong way to restore soil health. The important thing is to be sure your approach is restoring, not degrading, the health of your soil.

Merriam-Webster defines a problem as "something that is difficult to deal with: something that is a source of trouble, worry, etc." There are many things in agriculture that seem to be problems: soil erosion, weeds, insect pests, crop diseases, drought, floods, salinity, lack of soil fertility, etc. But to truly identify and understand a problem, I believe you must ask the question, "Why?" until there are no more answers.

Allow me to use soil erosion as an example. Soil is leaving a field carried by water that is running off the soil surface. Why? Because the soil particles are becoming detached from the surface and moving with the water running off the field. Why? Because the water is not infiltrating into the soil. Why? Because there are no water-stable soil aggregates at the soil surface protected by plants or plant residues. Why? Because the soil has not been managed in a way to create and protect stable soil aggregates at the soil surface. Why? Because the land manager does not understand to how to manage the soil to foster and protect stable soil aggregates. Why? Because no one involved in the situation understood that a lack of stable soil aggregates at the soil surface was the reason the water was not infiltrating into the soil, resulting in water runoff and soil erosion. Everyone thought erosion was the problem and focused on dealing with the detached soil and water that ran off the field. Witness the miles of terraces, waterways, diversions, filter strips, and buffers, across our cropland landscape. These practices were installed with good intentions, but serve only as Band-Aids® and diapers, rather than solutions, to the problem of dysfunctional soil.

Soil erosion, excess nutrients in runoff water, fish-kills, sedimentation, floods, droughts, weeds, insect pests and disease outbreaks can be attributed, all or in part, to dysfunctional soil. *Dysfunctional soil is the problem in agriculture that needs to be solved.* There is very little fresh water or plant nutrient that does not, at some point, move through or interact with the soil.

We must proceed with caution when addressing symptoms and applying tools in agriculture so we do not lose sight of the goal of restoring soil health. With fully functioning soil as our goal, we can begin to apply tools thoughtfully and purposefully to move toward that goal. Monitoring the capacity of the soil to function along the way will help determine what activities are most effective at restoring soil health. Many different approaches may be taken and many different tools may be used. An approach that works in one region may not work in another region. Capital, labor, education and technology will all play a role in what each producer chooses to do with his or her land to restore soil health.

So beware! There is no implement, additive, amendment, or product that will restore soil health for you. *Only an understanding of how the soil functions as a biological system can restore soil health.* The path you choose and the tools you use to achieve the goal of fully functioning soil on your farm, ranch or in your garden is up to you.

CHAPTER 7

How Will I Know
If My Soil Health
is Improving?

Monitoring soil health is not particularly difficult or expensive and can often be accomplished with some thoughtful observation and a shovel. Soil health assessment ranges from simple inexpensive tests you can perform yourself to more elaborate tests that you can have performed by laboratories in the U.S. for a fee.

The United States Department of Agriculture Natural Resources Conservation Service's (NRCS) soil health website has extensive information on how to assess soil health. See the appendix for links to that information. There is information on soil health indicators and how to assess them. There is also videos, guides and worksheets that you can watch, read and use to learn more about soil health.

Assessing soil health is not difficult. You can go out into your field or garden right now with a shovel and dig a hole a foot or so deep and carefully examine the soil. As you do so, examine the color, consistency, aroma, organisms, and the roots of the plants growing in the soil. Compare what you find to the soil health assessment information from NRCS or Cornell University (see appendix). Note that the terms "soil quality" and "soil health" have been used interchangeably over the years, so don't let that confuse you.

A few tests that I prefer to use to assess soil health are to smell the aroma of the soil, slake some dry soil aggregates in water, and measure the rate water infiltrates the soil. These tests are described in the NRCS and Cornell soil health assessment information, but I will cover them here briefly, so you can begin to do them now if you wish.

As mentioned in Chapter 3, a healthy soil should smell slightly sweet with the distinct earthy aroma of geosmin, a by-product of actinobacteria. A soil that smells like rotten eggs (hydrogen sulfide) denotes a soil dominated by anaerobic organisms. Healthy soil should instead be aerobic – containing oxygen. A soil with a metallic or kitchen-cleanser-bleach aroma is often dominated by bacteria and out of biological balance. A soil that has no detectable aroma is either dry, or has very few active soil organisms present. If the soil is dry, put some water on it and give the organisms I chance to reactivate, then check for aroma again a short time later. To capture the aroma of the soil, take a fresh sample of soil in your hands, crumble it slightly to break open the aggregates and immediately bring the soil to your nose and sniff.

A simplified soil slake test (details for a more exacting approach to this test is in the NRCS Soil Quality Test Kit Guide) can be performed by collecting some soil aggregates from the surface of the soil, allowing them to dry and then placing them in water to see how long they hold together. Collect some lumps, clumps or crumbs of soil from the soil surface, none larger than a golf ball. Place them in the sun, or inside on a workbench, for a day or two to allow the aggregates to become thoroughly dry. Once dry, immerse the aggregates in water (a clear jar works well for this) and observe how long the aggregates remain intact as the water seeps into the aggregate and the air bubbles out. It is best to gently drop the aggregates into a jar of water rather than pour water over the aggregates. If your soil aggregate disintegrates in less than a minute you have some work to do to improve the health of your soil! Remember, aggregates that disintegrate quickly will seal the soil surface quickly when it rains or irrigation water is applied, not allowing water or air to infiltrate the soil. Aggregates that are stable when wet allow water, air, roots, and soil organisms to move into and throughout the soil. Water-stable soil aggregates will remain intact while immersed in water for days or weeks without appreciable disintegration.

The third simple test is the water infiltration test. This test involves pressing or pounding a ring of some sort into the soil surface then pouring some water inside the ring to observe how long it takes for the water to infiltrate into the soil. A metal can with the top and bottom removed may suffice for this test. Drive the ring at least three inches into the soil. Add water to an inch in depth and begin timing how long it takes for that inch of water to disappear into the soil. After the first inch of water is gone, repeat the process with a second inch of water. Observing the infiltration of a second inch of water will give you a good appreciation of how the soil will infiltrate water after the surface of the soil is thoroughly wet. Rapid water infiltration is good, slow water infiltration is not. If the second inch of water takes more than 30 minutes to infiltrate into the soil, there is significant room for soil health improvement. Details on how to conduct this test with precision is included in the NRCS Soil Quality Test Kit Guide.

Performing these three fairly simple soil health assessments should give you a general idea of the health of your soil. Do not be alarmed or dismayed if your soil performs poorly in these tests. As mentioned earlier, most agricultural soils have been significantly, if not severely, degraded from their native condition. With the information included in this book, you will be able to educate yourself and work to improve the health of your soil.

There are some relatively new soil testing methods that focus on the biology of the soil, to give you a more detailed picture of the organisms that live there, and their nutrient cycling capabilities. The Haney method of soil analysis combines a relatively passive water extraction of nutrients from the soil with a measure of the soil organisms and their food supply. This method seeks to let the soil reveal what it needs, rather than simply reducing the soil to its chemical constituents. The test reveals the biologically derived plant nutrients based on the capacity of soil organisms to cycle nutrients. Tests are also available that reveal the populations and diversity of organisms in the soil, using various biologically produced compounds as indicators. These tests can estimate what types of organisms are present in your soil, as well as their numbers. The tests may also include an estimate of the amount of crop nutrients that will become available in your soil. See the appendix for labs that can perform various types of soil health related tests.

An example of the economic benefit of improving soil health, is in a recent revision to the North Dakota State University soil test recommendations by Dr. Dave Franzen, Extension Soils Specialist. The notable change is that the nitrogen recommendation calculations now provide for a 50 pound nitrogen credit on cropland that has not been tilled for at least five consecutive years. From extensive data review and research, Dr. Franzen has determined that soils under a system of crop production without tillage will cycle nitrogen differently than soils that are tilled on an annual basis. His nitrogen recommendation calculator recognizes that during the transition from a tilled system to a no-tillage system, the soil should have slightly more (20 pounds per acre) nitrogen applied on an annual basis to help the soil microbes break down any high C:N ratio crop residues to establish a new equilibrium in the soil. After five years of continuous no-tillage, the soil achieves a new balance and a higher level of biological function which actually supplies more nitrogen to the growing crop (50 pounds per acre per year) than a similar soil that is annually tilled.

The 50 pound per acre per year nitrogen credit is solely a function of no-tillage of the soil while producing annual crops. The lack of soil disturbance provides a stable habitat for soil microorganisms to thrive, increasing soil organic matter and the capacity of the soil to supply nitrogen to growing crops. Producers practicing no-tillage methods have realized the economic benefit of the 50 pound nitrogen credit. Dr. Franzen's wheat nitrogen calculator can be accessed at: https://www.ndsu.edu/pubweb/soils/wheat/

Now It Is Up to You!

I now encourage you to become a student of soil health. You may need to read this book more than once and allow yourself some time for the concepts to become clear to you. Armed with the information included in this book, you can increase your knowledge of how the soil functions and restore and assess soil health. Choose a part of your farm, ranch or garden to apply the principles of improving soil health until you demonstrate to yourself how the soil will change for the better. Then you can confidently apply the principles across more of your land.

Perform some of the soil health assessments on your soil to see where it stands in its capacity to function. If the assessment shows that your soil health is not what you would like it to be, do not be discouraged. Begin applying the principles for improving soil health and monitor your soil to see how your efforts are making a difference. You now have the knowledge to begin to restore your soil. How and when you choose to apply this knowledge is up to you. I wish you great success!

Appendix

United States Department of Agriculture Natural Resources Conservation Service
http://www.nrcs.usda.gov/wps/portal/nrcs/main/national/soils/health/ and
http://www.nrcs.usda.gov/wps/portal/nrcs/main/soils/health/assessment/

Brown's Ranch
http://brownsranch.us/

Dakota Lakes Research Farm
http://www.dakotalakes.com/

No-till on the Plains
http://www.notill.org/

Burleigh County Soil Conservation District
http://www.bcscd.com

USDA NRCS Science & Technology Training Library Webinar Portal for Conservation of Natural Resources
http://www.conservationwebinars.net/

Green Cover Seed
https://greencoverseed.com/

Cornell University, College of Agriculture and Life Sciences
Cornell Comprehensive Assessment of Soil Health – The Cornell Framework
Manual

Carbon:Nitrogen Ratios

C:N	Material
80:1	Wheat straw
70:1	Oat straw
57:1	Corn stover
37:1	Rye, flowering
29:1	Soybean residue
26:1	Rye, fresh before flowering
25:1	Alfalfa hay
24:1	**Ideal microbial diet**
23:1	Clover, fresh
20:1	Cow manure, rotted
19:1	Mustard, fresh
19:1	Turnip, fresh
18:1	Broccoli, fresh
18:1	Cauliflower, fresh
17:1	Grass clippings, fresh
11:1	Hairy Vetch, fresh
8:1	Cow manure, fresh
2:1	Root exudates

Materials with a C:N ratio **greater than 24:1** will make nitrogen less available for plants.

Materials with a C:N ratio **less than 24:1** will make nitrogen more available for plants.

Materials with a C:N ratio **greater than 24:1** will decompose more slowly in the soil.

Materials with a C:N ratio **less than 24:1** will decompose more rapidly in the soil.

Soil Testing Services to Assess Soil Health (for informational purposes only, not an endorsement)

- Earthfort. Earthfort.com. Soil Food Web analysis to show the type and numbers of microscopic organisms living in your soil including nitrogen cycling potential.

- Ward Laboratories. Wardlab.com. Soil health/biotesting. Phospholipid Fatty Acid (PLFA) analysis to describe the soil microbial community. Haney test, which integrates both chemical and biological testing methods to estimate nutrient availability in the soil from a soil food web and plant perspective. Solvita test to estimate active soil carbon and microbial biomass.

Glossary

Aerobic – Requiring oxygen, typically from the air, but sometimes from oxygen dissolved in water.

Aggregates – lumps, clumps and crumbs of soil made up of sand, silt and clay stuck together by sticky substances produced by plants and organisms that live in the soil.

Anaerobic – Not requiring oxygen.

Arbuscular Mycorrhizae Fungus – a type of fungus that penetrates the roots of plants and creates a mutually beneficial relationship between the fungus and the plant. The plant provides sugars (food) to the fungi and the fungi brings water, phosphorus and other substances from the soil to feed the plant.

Biodiversity – a contraction of "biological diversity". The array of different species in a given place, such as the soil.

Biogeochemical – denoting that a substance may undergo changes from interactions with living organisms (bio), the mineral component of the soil (geo) and reactions that take place in water solutions in the soil (chemical).

Biome – the collection of plants and animals that occupies a given place on Earth, such as the soil.

Buffering – the capacity to prevent large changes in acidity or alkalinity.

Clay – soil particles smaller than silt that become plastic when wet and hard when dry.

Colloids – a condition where particles (typically clay) are dispersed in a water solution in the soil and remain suspended in the water for an extended period of time.

Copiotrophic – organisms found in environments rich in nutrients such as nitrogen and carbon that feed and reproduce quickly.

Cycling – changing the form and location of an element (i.e. nitrogen) as it moves through the soil, air, water, plants or soil organisms.

Detritusphere – a layer of plant material that covers the soil.

Exudates – substances that flow slowly out of plant roots, often consisting of amino acids and sugars.

Exude – to flow out slowly, as sugary substances from plant roots.

Filtering – Separating or removing a substance from water as it passes through the soil.

Geosmin – an organic compound with a slightly sweet, earthy aroma, produced by Actinobacteria.

Glomalin – a sugar and protein substance produced abundantly on hyphae and spores of arbuscular mycorrhizal fungi in soil and in roots.

Hyphae – thread-like portion of a fungus that can access very small spaces in the soil.

Humus – organic matter that has decomposed to a stable form in the soil. Usually black or dark brown in color.

Partitioning – Dividing something into two or more quantities.

Photosynthesis – the process that takes place in a green plant turning water and carbon dioxide into food when the plant is exposed to light.

Polysaccharide – a combination of two or more simple sugar (glucose) molecules linked into a simple or branched chain.

Sand – a loose granular material that results from the disintegration of rocks and consists of particles smaller than gravel but coarser than silt.

Separates – The individual particles of sand, silt and clay.

Silt – a loose sedimentary material that consists of particles smaller than sand but larger than clay.

Slake - Slaking is the disintegration of air-dry soil aggregates into smaller sized aggregates or soil separates when they are suddenly immersed in water.

Solute – A dissolved substance, particularly dissolved in water.

Symbiotic – a cooperative relationship between different organisms such as a plant and a fungus.

Treatise – A book that discusses a subject carefully and thoroughly.

Bibliography

Bardgett, R. and W. van der Putten. 2014. Belowground Biodiversity and Ecosystem Functioning. Nature. Vol. 515. Pp. 505-511.

Bauer, A. and T. Conlon. 1974. Influence of Tillage Interval of Fallow on Soil Water Storage. North Dakota Agricultural Experiment Station.

Beare, M. H., et al. 1995. The Significance and Regulation of Soil Biodiversity. Springer Netherlands. 5-22.

Beck, D. 2005. The Power Behind Crop Rotations. Dakota Lakes Research Farm, Pierre, SD.

Brady, Nyle C. and Ray R. Weil. 2002. The Nature and Properties of Soils. 13th Ed. Pearson Education Inc. pp. 176-177.

Culliney, T. 2013. Role of Arthropods in Maintaining Soil Fertility. Agriculture. Vol. 3. Pp. 629-659.

Davidson, E. and I. Janssens. 2006. Temperature Sensitivity of Soil Carbon Decomposition and Feedbacks to Climate Change. Nature. Vol. 440. Pp. 165-173

Doran, J.W., and T.B. Parkin. 1994. Defining and assessing soil quality. p. 3-21. In J.W. Doran et al. (ed.) Defining soil quality for a sustainable environment. SSSA Spec. Publ. 35. SSSA and ASA, Madison, WI.

Drinkwater, L.E. & S.S. Snapp. 2007. Nutrients in agriculture: Rethinking the management paradigm. Advances in Agronomy. 92:163-186.

Edwards, C.A. and P.J. Bohlen. 1977. Biology and Ecology of Earthworms. Chapman and Hall.

Eisenbraun, T. 2011. Southwest North Dakota Soil Health Demonstration project final report. Project LNC09-312. www.http://mysare.sare.org/MySare/ProjectReport.aspx?do=viewRept&pn=LNC09-312&y=2011&t=1

Ellsbury. 2009. Organic Matter and Water Stability of Field Aggregates Affected by Tillage in South Dakota. Soil Science Society of America Journal. Vol. 73 No. 1, p. 197-206.

Hatfield, J. 2016. The Science Supporting Changes in Soil Health. Science and Technology Training Library Webinar Portal for Conservation of Natural Resources.

http://www.conservationwebinars.net/webinars/the-science-supporting-changes-in-soil-health/

Hoorman, J. and R. Islam. 2010. Understanding Soil Microbes and Nutrient Cycling. Ohio State University Extension and Midwest Cover Crops Council. Fact Sheet SAG-16.

Hudson, B. 1994. Soil Organic Matter and Available Water Capacity. Journal of Soil and Water Conservation. Vol. 49, No.2, Pp.189-194

Jones, C. 2008. Liquid Carbon Pathway. Australian Farm Journal. Edition 338.

Laboski, C.A., J.E. Sawyer, D.T. Walters, L.G. Bundy, R.G. Hoeft, G.W. Randall, and T.W. Andraski. 2006. Evaluation of the Illinois Soil Nitrogen Test in the North Central Region. North Central Extension-Industry Soil Fertility Conference. November 7-8, 2006. Des Moines, IA. Vol. 22, p. 86-93.

Lavelle, P. and Alister V. Spain. 2005. Soil Ecology. Kluwer Academic Publishers. p.112, pp.430-440.

Lonhienne, Chanyarat P., Thierry G. A. Lonhienne, Doris Rentsch, Nicole Robinson, Michael Christie, Richard I. Webb, Harshi K. Gamage, Bernard J. Carroll, Peer M. Schenk, and Susanne Schmidt. 2008. Plants Can Use Protein as a Nitrogen Source Without Assistance From Other Organisms. Proceedings of the National Academy of Sciences 2008, Vol. 105, No. 11, pp. 4524 – 4529.

Lowdermilk, W. 1953. Conquest of the Land Through 7,000 Years. USDA Washington, D.C.

Manitoba Agriculture, Food and Rural Initiatives. 2008. The Basics and Benefits of Bale Grazing. www.gov.mb.ca/agriculture/livestock/production/beef/pubs/baa05s04j.pdf

Montgomery, D. 2007. Dirt: The Erosion of Civilization. University of California Press.

Nardi, J. 2003. The World Beneath Our Feet A Guide to Life in the Soil. Oxford Press. Pp. 72-159.

Nichols, K. 2008. Glomalin What is it and what does it do? Agricultural Research July 2008. Pp. 20-21.

Pikul, JL; Chilom, G; Rice, J; Eynard, A; Schumacher, TE; Nichols, K; Johnson, JMF; Wright, S; Caesar, T; Ellsbury, M. 2009. Organic Matter and Water Stability of Field Aggregates Affected by Tillage in South Dakota. Soil Science Society of America Journal. 73(1):197-206.

Reicosky, D.C. and M.J. Lindstrom. 1993. Fall Tillage Method: Effect on Short-Term Carbon Dioxide Flux from Soil. Agronomy Journal Vo. 85. Pp. 1237-1243.

Savory, A. 1988. Holistic Resource Management. Island Press. P. 108.

Smith, Sally E., Iver Jakobsen, Mette Grønlund, and F. Andrew Smith. 2011. Roles of Arbuscular Mycorrhizas in Plant Phosphorus Nutrition: Interactions between Pathways of Phosphorus Uptake in Arbuscular Mycorrhizal Roots Have Important Implications for Understanding and Manipulating Plant Phosphorus Acquisition 1. Plant Physiology, July 2011, Vol. 156, pp. 1050–1057.

Stevens, W.B., Hoeft, R.G. and Mulvaney, R.L. 2005. Fate of Nitrogen-15 in a Long-Term Nitrogen Rate Study. Agronomy Journal Vol. 97, No. 4. Pp. 1046-1053.

Tilman, D., P. Reich & J. Knops. 2006. Biodiversity and Ecosystem Stability in a Decade-Long Grassland Experiment. Nature. Vol. 441. Pp. 629-632.

Van Meter, K J, N B Basu, J J Veenstra and C L Burras. 2016. The Nitrogen Legacy: Emerging Evidence of Nitrogen Accumulation in Anthropogenic Landscapes. Environmental Research Letters, Vol. 11, No. 3.

Walker, T., H. Bais, E. Grotewold, and J. Vivanco. 2003. Root Exudation and Rhizosphere Biology. Plant Physiol. Vol. 132, pp. 44-51.

Wetterauer, D. & J. Killorn. 2013. Fallow and Flooded Soil Syndromes: Effects on Crop Production. Journal of Production Agriculture. Vol. 9 No. 1, p. 39-41.

Zak, Donald R., William E. Holmes, David C. White, Aaron D. Peacock and David Tilman. 2003. Plant Diversity, Soil Microbial Communities and Ecosystem Function: Are There Any Links? Ecology August 2003, Vol. 84, pp. 2042–2050.